INTERNAL-COMBUSTION

LOCOMOTIVES AND MOTOR COACHES

BY

† PROF. I. FRANCO

MECHANICAL ENGINEER,
LATE PROFESSOR AT THE TECHNICAL HIGH SCHOOL OF DELFT (HOLLAND),
LATE SUPERINTENDENT OF MOTIVE POWER, ELECTRIC TRACTION AND
ROLLING STOCK DEPARTMENT OF THE NETHERLANDS RAILWAYS

AND

P. LABRYN

MECHANICAL ENGINEER,
CHIEF OF LOCOMOTIVE CONSTRUCTION DEPARTMENT
OF THE NETHERLANDS RAILWAYS

†

Springer-Science+Business Media, B.V. 1931

ISBN 978-94-017-5765-2   ISBN 978-94-017-6155-0 (eBook)
DOI 10.1007/978-94-017-6155-0

PREFACE.

Though much has already been written on the subject of traction by means of locomotives and motor coaches driven by combustion engines, this has to be sought in a wide range of various periodicals, which fact induced the late Professor Franco to embody these scattered publications in a book for ready reference by railway engineers and managements. He considered it advisable to treat first and foremost of the use of combustion engines in rail-motor coaches, small locomotives and similar traction vehicles.

Professor Franco invited me to assist in the compilation of this work, and this invitation I gladly accepted.

It was decided that no special scientific treatise on combustion engines and accessories was required, but rather a book that could be consulted by anyone interested in the problem of the application of combustion engines on railways in order to see what had already been attained in this direction in the various countries of the world. Furthermore it was considered that an extensive bibliography would enable the reader to trace any further information he might desire.

Unfortunately Professor Franco, who was my tutor and afterwards my chief on the Netherlands Railways, died just as the manuscript reached completion, but nevertheless I believe to be acting in accordance with his wishes and those of his family in having this book published as the fruit of our joint labour.

P. LABRYN,
Mechanical Engineer.

Utrecht, April 1931.

4

CONTENTS.

ABBREVIATIONS.

atm.	= atmosphere(s)	in.	= inch(es)
atm.	= atmospheric	kcal	= kilogramme calorie
Ah	= ampere hour	kg	= kilogramme
B.T.U.	= British thermal unit	km	= kilometre
B.H.P.	= brake horse-power	kW	= kilowatt
C.	= Celsius	kWh	= kilowatt hour
C.P.	= candle power	lb(s).	= pound(s)
cap.	= capacity	loc.	= locomotive
circ.	= circumference	lub. oil	= lubricating oil
cm	= centimetre	m	= metre
cm²	= square centimetre	m²	= square metre
cons.	= consumption	m³	= cubic metre
cub.	= cubic	max.	= maximum
cyl.	= cylinder	min.	= minute
DC	= direct current	min.	= minimum
dia.	= diameter	mm	= millimetre
driv. wheel	= driving wheel	m.p.h.	= miles per hour
eff.	= efficiency	revs.	= revolutions
E.H.P.	= effective horse-power	sec	= second(s)
F.	= Fahrenheit	sq.	= square
ft.	= foot or feet	t	= ton(s)
g	= gramme	temp.	= temperature
glns.	= gallons	tkm	= ton kilometre
h	= hour	tract. force	= tractive force
H.P.	= horse-power	vol.	= volume
H.P.h	= horse-power hour	Wh	= watt hour
I.H.P.	= indicated horse-power		

———

INTRODUCTION.

Transport is one of the most important problems of the present day. Notwithstanding the increase of motor transport during recent years and the continuous advance in air transport, railway transport will always occupy a very prominent place.

Since the beginning of railways more than a century ago the only traction vehicle until comparatively recent date was the steam engine, and this, through years of experience, has reached a high standard of perfection.

Nevertheless the efficiency of the most perfect steam locomotive is still very small. Various attempts have been made to increase its efficiency by applying turbines or steam under very high pressure, but so far very little has been achieved in this direction beyond the experimental stage, so that more general application cannot yet be expected. Thus it is not at all surprising that another way should have been sought to build a more economical locomotive. This led to the use of internal-combustion engines instead of steam engines, a rather obvious solution considering the great success already attained with this type of engine on board ships. Although the requirements of marine engines are different and in some ways less exacting, the still greater perfection reached in the design of internal-combustion engines and the improvements which are constantly being made as more experience is gained, justify the expectation that this engine will also give every satisfaction in railway work. It may even be said now already that in special cases the internal-combustion engine is the one most eminently suitable for locomotives, as indeed appears from the fact that railway companies all the world over have taken to this type.

A great advantage of internal-combustion engines is that they can be applied not only to locomotives but also to motor vehicles, a possibility which is practically precluded with steam engines; at any rate they have not been generally used in this way. A comparison of the advantages and disadvantages of the steam engine with those of the internal-combustion engine as applied to traction vehicles shows that:

The advantages of the internal-combustion engine compared with the steam engine are:

1) Much higher efficiency.

This efficiency, that is to say the proportion of the fuel energy converted into work performed at the periphery of the driving-wheels, is no more than 8% for the most perfect steam locomotive, whereas with the internal-combustion engine locomotive an efficiency of 25% is reached.

2) Readiness for immediate service.

In a steam locomotive the steam has first to be generated, which, unless the engine is kept under steam, occupies a few hours and means extra cost of fuel and wages.

3) No fuel costs while the locomotive is at a standstill.

With steam engines the fire has to be stoked whether the locomotive is in motion or not.

4) In some cases there is no cost of wages while the locomotive is idle.

For shunting engines, which, especially at smaller stations, are idle for several hours a day, a saving is effected through the engine-driver being able to do other work or take his off-hours during that time.

5) No consumption of water.

This is especially an advantage in countries where water is scarce, or where boiler-feed water is expensive.

6) The possibility for longer journeys.

After a journey of 200 to 300 km the fire of a steam engine has to be cleaned. An internal-combustion engine can be kept running for any length of time.

7) Generally a lower wage-bill.

A steam train requires at least an engine-driver, a fireman and a guard.

A motor train needs only one man. For a motor shunting engine one man suffices, whereas a steam shunting engine needs an engine-driver and a fireman.

8) Cheaper handling of the fuel at depôts.

The liquid fuel only requires pumping out of the tank cars into the storage tanks and out of these into the tank of the traction vehicle.

The disadvantages are:

1) Higher initial cost.

A motor locomotive to-day costs about twice as much as a steam locomotive. For motor vehicles the cost has to be compared with that of a small steam locomotive plus trailer, and in this case the difference in price will be of little importance.

2) Smaller overload-capacity.

An internal-combustion engine can only be slightly overloaded and only during a short time, as otherwise troubles may occur through fouling of the engine. A steam engine, on the other hand, can be considerably overloaded, though at the cost of efficiency.

3) More complicated construction.

4) Heavier weight for locomotives.

The data given hereafter for locomotives already built will give some idea of this factor.

Another factor is, of course, the cost of fuel, but at the time of writing this book fuel prices have changed to such a considerable extent that it is deemed inadvisable to make any comparisons. When considering the cost due thought is also to be given to local conditions.

I. THE ENGINE.

Types.

Diesel and petrol engines are generally called internal-combustion engines. Combustion engines are engines consisting of one or more cylinders in each of which a piston moves reciprocatingly. This movement of the piston is converted by means of a connecting-rod and a crank into a rotatory motion. The piston is propelled by the ignition of the fuel fed into the cylinder, the combustion gas enclosed between cylinder cover and piston exercising a strong pressure upon the piston and thus setting it in motion, while the gas gradually expands. In this respect there is much similarity between the combustion engine and an ordinary steam engine, the typical difference being that the combustion gas inside the cylinder of the combustion engine has a much higher temperature than the steam in a steam engine. This makes it necessary to provide the cylinders of combustion engines with cooling jackets. Partly in consequence of this intense heat double-acting engines are apt to give trouble at the stuffing-box and with the cooling of the piston, and for this reason combustion engines are nearly always made single-acting, except in the case of the very largest types, which, however, up to the present are not to be considered for traction purposes.

There are two principal types of combustion engines: the Diesel engine and the explosion engine. Formerly the Diesel engine was also called a constant pressure engine, as the combustion was rather gradual, the explosion engine also a constant volume engine, as the combustion was rather instantaneous. To-day this contrast has disappeared as both types of engines have a more mixed type of combustion [1].

Figures 1 and 2 show the theoretical diagrams of a constant volume engine and a constant pressure engine, whilst figure 3 shows a theoretical diagram of a more mixed type of combustion. Figures 4 to 9 are illustrations of various constructions, from which it will be seen that these engines when used for locomotives are invariably of the vertical type.

The combustion cycle.

To attain a high efficiency it is necessary that the fuel charge burns completely. Complete combustion can only be effected with a surplus of air, because only one-fifth part of the air consists of oxygen, making it less effective. This surplus of air is particularly needed also because the combustion has to be completed quickly and consequently every particle of the fuel must receive immediately the necessary quantity of oxygen. For the combustion of one kg of fuel about 15 m^3 air is required in a petrol engine and about 20 m^3 in a Diesel engine.

---

[1] The characterising feature of a Diesel engine, according to the inventor, is: Compression of the combustion air without fuel until considerably above the ignition point of the fuel; then the adding of the fuel and the immediate combustion of same in this highly compressed hot air. If, in an internal-combustion engine, air without fuel is compressed but instead of the ignition taking place through the hot compressed air the fuel is ignited by other means (for instance by a hot surface) then this engine cannot really be called a Diesel engine.

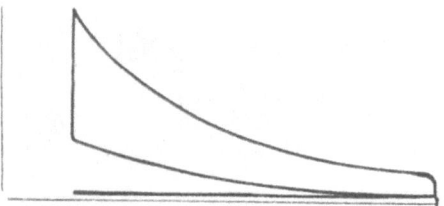

Fig. 1. Indicator-diagram of Explosion-engine.

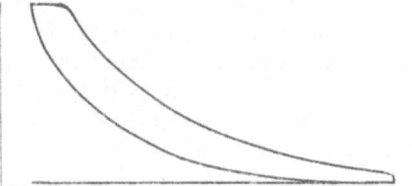

Fig. 2. Indicator-diagram of Constant-pressure engine.

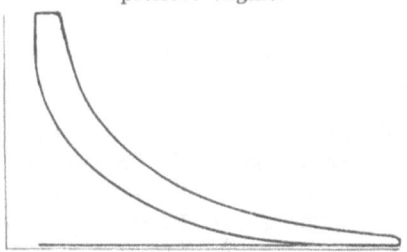

Fig. 3. Indicator-diagram of mixed type of combustion.

Fig. 4. Kerr Stuart Diesel engine 30 H.P., 800 revs. p. min, cylinder bore 135 mm, stroke 200 mm.

The highest temperature of combustion in petrol engines may be estimated at roughly 4000°—4500° F. and in Diesel engines 2500°—3500° F.; with lower loads of the engine the temperature is reduced.

Number of revolutions.

Explosion engines can be built to make a greater number of revolutions than Diesel engines. This is an important advantage of explosion engines because they can be built lighter and cheaper for the same power. On the other hand the cost of the fuel is also an important factor.

Fuel.

Petrol and Diesel engine fuel are both petroleum products. For Diesel engines kerosene, gasoil and heavy oils, so-called Diesel oils, can be used.

All these liquids are distilled from mineral oil in different fractions, so that their boiling points vary.

Boiling temperature.

Petrol   120—400° Fahrenheit.
Kerosene   320—570   „
Gasoil & Diesel oil   400—700   „   and higher.

(see graph 10, p. 19)

Fig. 5. Ingersoll-Rand, six cylinder, four-cycle, single-acting, solid injection, Diesel engine of 750 H.P.

Fig. 6. Solid injection, six cylinder, M.A.N. Diesel engine, 600 H.P., Bore 280 mm, Stroke 380 mm, 700 revs. p. min.

The calorific value of all these fuels is about 10000 kcal per kg (18000 B.T.U./lb.). When determining the calorific value of fuels to be used in an engine it is really the calorific value of the mixture of air and fuel contained in the engine that has to be considered, because this fixes the quantity of heat released by every cylinder charge. This also decides the amount of work done and thus also the size of cylinder for the required power. As a matter of fact one m^3 of mixture contains about 500—650 kcal.

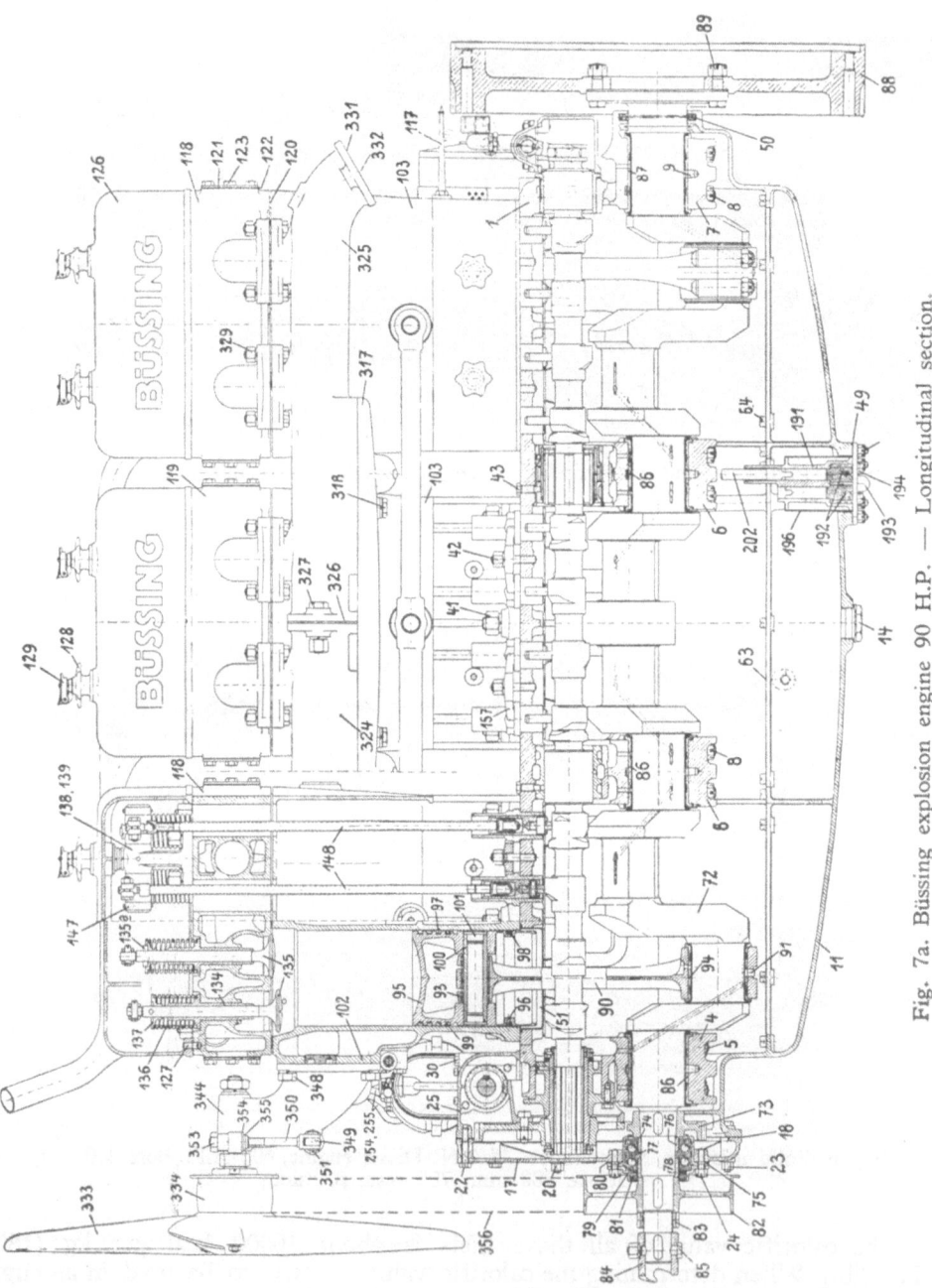

Fig. 7a. Büssing explosion engine 90 H.P. — Longitudinal section.

Kerosene does not lend itself readily for use in explosion engines because of its high boiling point (390° F. at atmospheric pressure), so that during the suction stroke it cannot be evaporated completely and consequently a good explosive mixture is not easily formed.

Fig. 7b. Side view of camshaft.

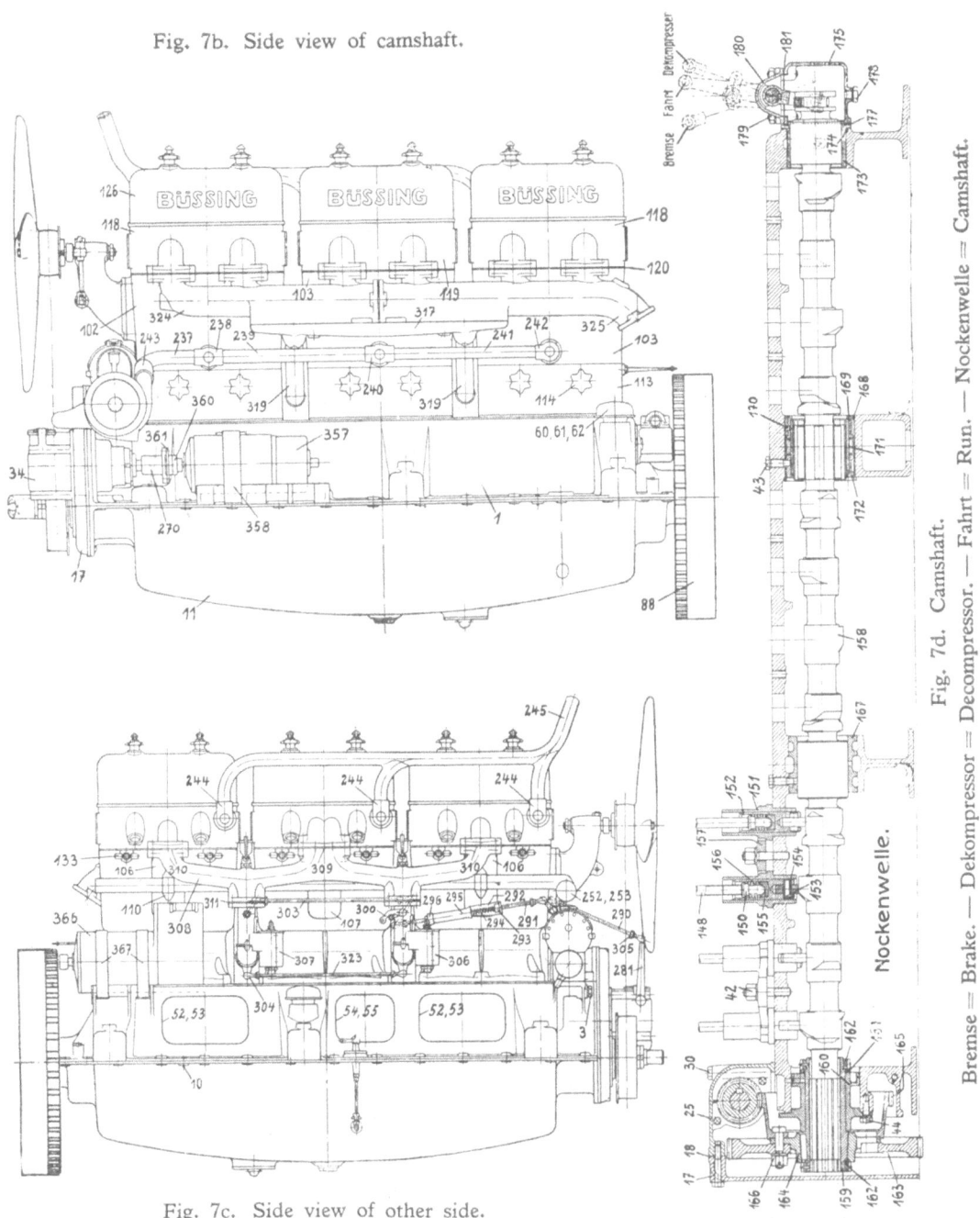

Fig. 7c. Side view of other side.

Fig. 7d. Camshaft.

Bremse = Brake. — Dekompressor = Decompressor. — Fahrt = Run. — Nockenwelle = Camshaft.

In order, however, that kerosene can be used in explosion engines, special carburettors are provided, by which the kerosene is heated up in contact with the air, in which state it undergoes a so-called pre-evaporation and pre-combustion. For starting, therefore, petrol is required.

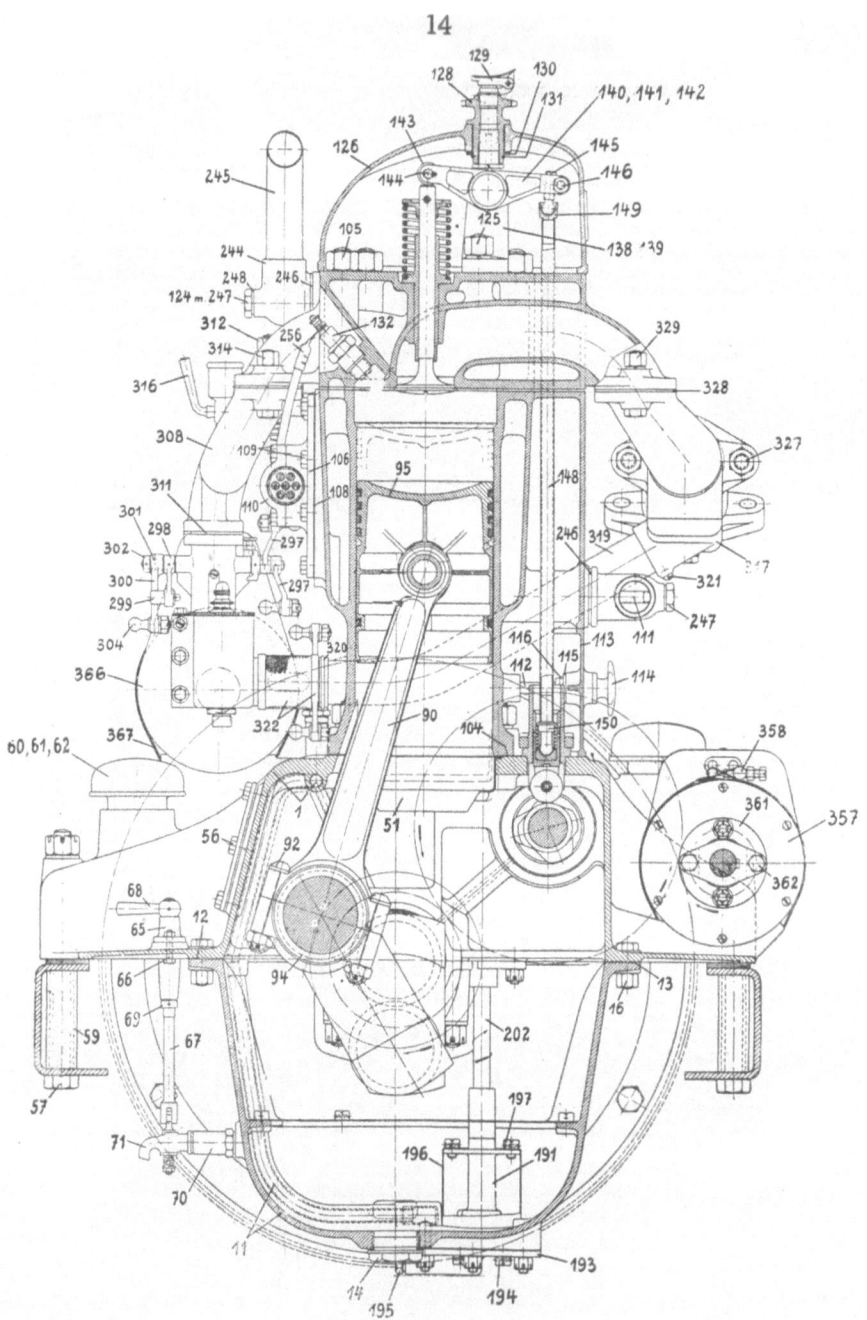

Fig. 7e. Cross section.

The need of heating the fuel and air mixture in kerosene engines makes it necessary to reduce the compression ratio, which means lower efficiency, and when considering the cost of running an engine from the point of view of fuel price this lower
efficiency has to be taken into account.

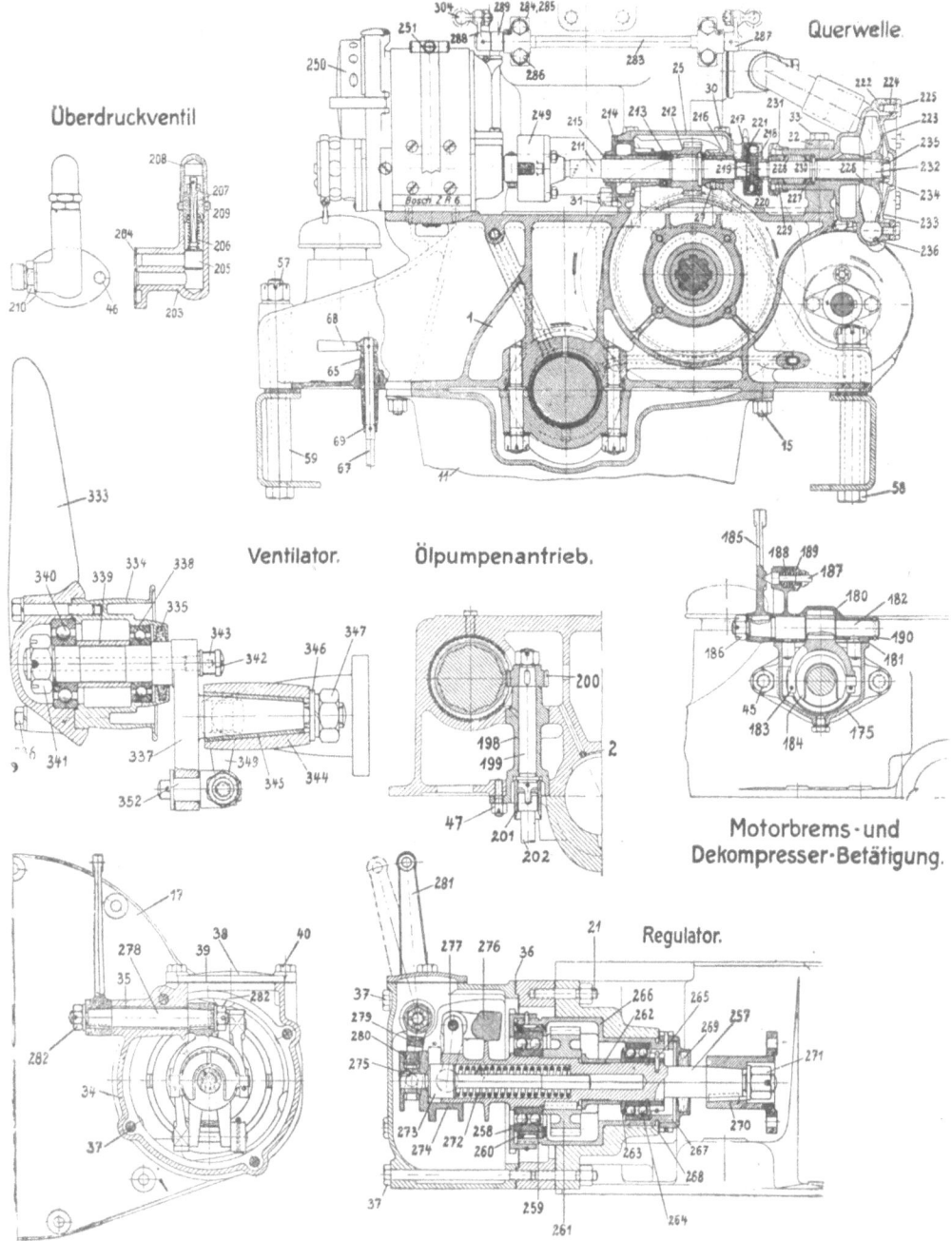

Überdruckventil

Querwelle.

Ventilator.

Ölpumpenantrieb.

Motorbrems- und
Dekompresser-Betätigung.

Regulator.

Fig. 7f.
Motorbrems- und Dekompressor-Betätigung = Engine brake and decompressor gear. — Ölpum-
penantrieb = Oilpump drive. — Querwelle = Intermediate shaft. — Regulator = Governor. —
Überdruckventil = Overflow-valve. — Ventilator = Fan.

Owing to its complicated construction the engine is one of the mechanical parts of a traction vehicle most subject to running troubles. In order to avoid break-downs it is therefore necessary that the various parts of the engine, especially those subjected to great forces, are amply dimensioned. Closely related to this is the

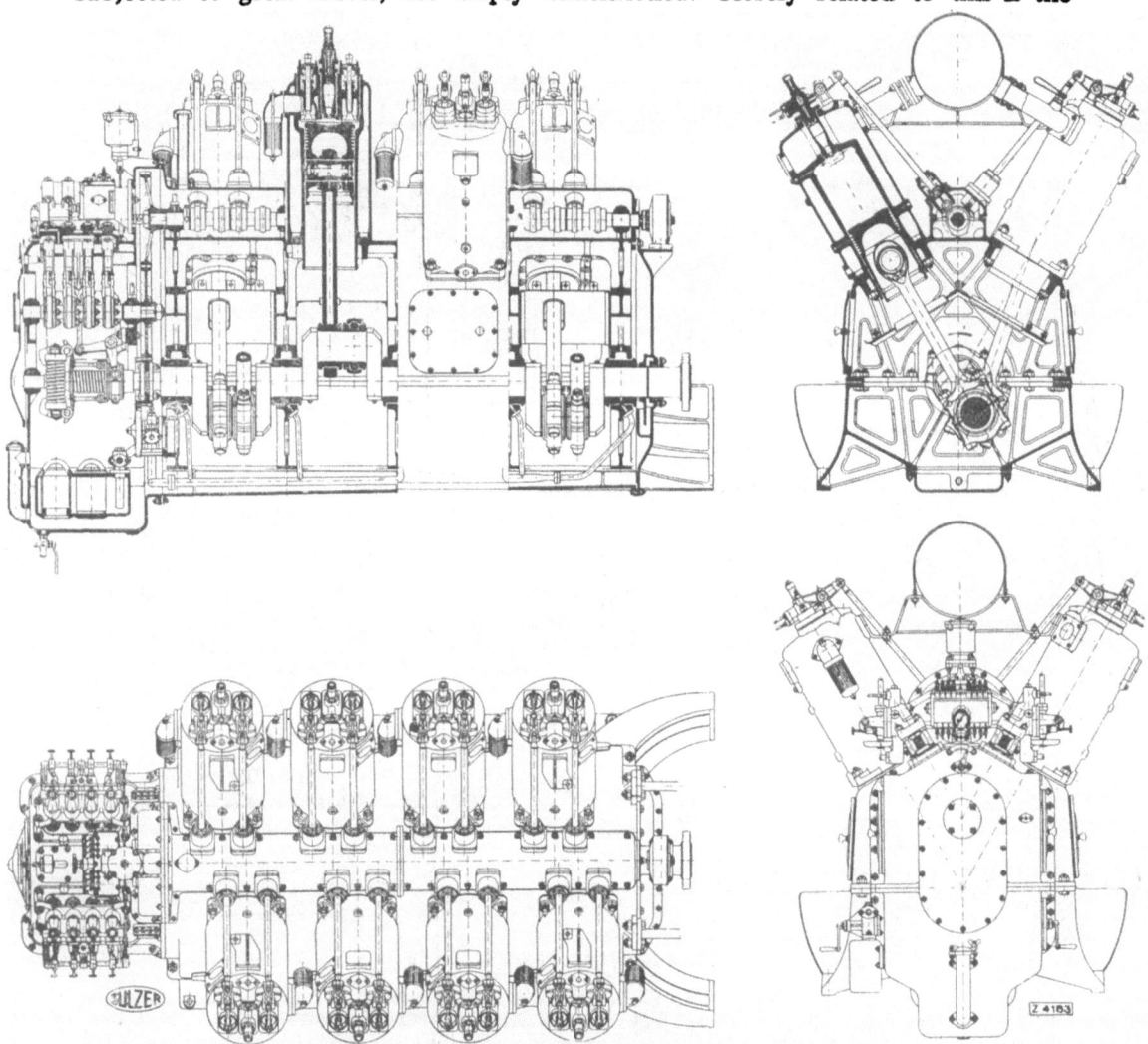

Fig. 8. Eight-cylinder Sulzer Diesel engine 250 H.P.

question how far compression can be increased. On the one hand a higher compression gives a higher efficiency, but on the other hand the forces acting on certain parts of the engine are considerably increased. For instance if in a petrol engine the compression is raised from 75 lbs. to 100 lbs. per sq. inch the efficiency will be from 12% to 15% higher; the carburettor, ignition and cooling have to be regulated accordingly. This increase of efficiency may mean a higher power or a lower fuel consumption, but the increase of power is not in proportion to the increase of efficiency, being in fact much smaller, because a higher compression means greater internal losses.

Benzol is also used as fuel for explosion engines; this is a product obtained from the distillation of coal. The advantage of benzol compared with petrol is that as a result of its higher ignition temperature it allows of a higher compression in the engine and thus increases the efficiency.

The calorific value of benzol, however, is about 10% less than that of petrol, so that the higher efficiency obtainable with benzol does not mean a much lower consumption.

Fig. 9. 60 H.P. Four cylinder Kerr Stuart Locomotive engine.

The properties of gasoil rendering it unsuitable for use in petrol engines make it, on the other hand, highly suitable for use in Diesel engines, where the fuel is added to intensively heated compressed air, in which spontaneous ignition takes place. In this case the low ignition temperature is favourable for spontaneous ignition and smooth combustion. For spontaneous ignition of the fuel, however, a higher compression is required than is usual in petrol engines, and this causes a higher pressure during the combustion, in consequence of which the rods and crankshaft are more heavily loaded. The maximum pressure in petrol engines generally amounts to 400 lbs. per sq. inch (sometimes more), whereas high speed Diesel engines may have a pressure of 600 lbs. per sq. inch, and for this reason the weight of a Diesel engine will be about 10—20% higher than that of a petrol engine with the same cylinder-volume, number of revolutions and horse power, the mean pressure in both types of engines being about the same. As a matter of fact the aim is to attain the same speeds in Diesel engines as in explosion engines, and in this direction much has already been achieved, as will be seen from the examples mentioned hereafter.

An advantage with gasoil as compared with petrol is that it is much less inflammable, so that there is less risk of fire.

The cycle (four-stroke or two-stroke).

To ensure that the gases inside the cylinder perform their utmost work by expansion, there must be the highest possible pressure prior to combustion. Consequently before combustion begins there must be compression. This applies both to explosion and to Diesel engines.

The cycle is as follows:
introduction of the fresh charge,
compression of this charge,
combustion and expansion,
expulsion of the exhaust-gases.

After this has taken place the piston must have resumed its original position for the introduction of a fresh charge.

This cycle can take place in two ways:
first forward piston stroke: drawing in of the fresh charge;
first return stroke: compression;
second forward stroke: combustion, followed by expansion of the combustion gases;
second return stroke: expulsion of the gases.

Thus during four piston strokes (two revolutions) there is only one working-stroke.

Engines working in the above manner are called *four-stroke engines.*

Two-stroke engines have one working-stroke for every two piston strokes (one revolution). On the return stroke the cylinder charge is compressed and on the following forward stroke combustion and expansion of the gases takes place. In the interval between the completion of expansion and the beginning of compression the expulsion of the exhaust gases takes place, immediately followed by the entry of a fresh charge.

The power of the two-stroke engine is not twice that of the four-stroke engine under corresponding conditions. As the mean pressure in the cylinders and the mechanical efficiency are less in the two-stroke engine than in the four-stroke engine, the power of the former is as a rule no more than $1\frac{1}{2}$ times that of the latter.

Introduction of the combustion air.

In explosion engines the fuel is mixed with the combustion air prior to compression. With the compression the temperature rises, but it must not reach or even approach too close to the ignition temperature of the mixture, as otherwise the well-known detonation will take place. Thus the extent to which compression can be raised is determined by the ignition temperature and engine characteristics; the higher the compression the better the work. Therefore it is advisable to provide for sufficient heat being carried off during the compression. Towards the end of compression the ignition takes place by means of an electric spark.

With Diesel engines the fuel is injected at the end of compression. In this case only air is compressed and consequently there is no danger of pre-ignition, so that the compression can be raised higher than in explosion engines.

The air is compressed to such an extent (about 450 lbs. per sq. inch and higher) that the temperature rises above the ignition temperature of the mixture of fuel and air (to about 1300° F.) and on the liquid fuel then being sprayed into the cylinder in a very finely divided state — either by means of highly compressed air of about 800 lbs. per sq. inch pressure or by means of small high-pressure pumps (about 2000 lbs. per sq. inch) without air —, it will ignite spontaneously and burn. Fuel is injected during about one-tenth of the stroke. If the fuel is injected without air no compressor is needed, and such engines are therefore called compressorless or solid injection engines; these are simpler in design than those which need air for the injection of the fuel.

Of late solid injection engines have become more and more popular. The mechanical injection of the fuel does away with the need of an air compressor, thus simplifying the design of the engine and increasing the mechanical efficiency from about 75 to 85%. The fuel consumption per I.H.P., however, is a little higher than that with engines in which the fuel is injected with compressed air. On the other hand a great advantage with solid injection engines, especially in the case of Diesel engines for locomotives and motor vehicles, is the lighter weight.

Solid injection engines are made with two different methods of fuel feed, viz.:

A. Fuel is injected into and partly burned in a special chamber separate from the combustion space of the cylinder. In consequence of the resultant pressure the mixture of fuel and combustion gases is blown into the combustion space proper in the form of a very fine mist and there thoroughly

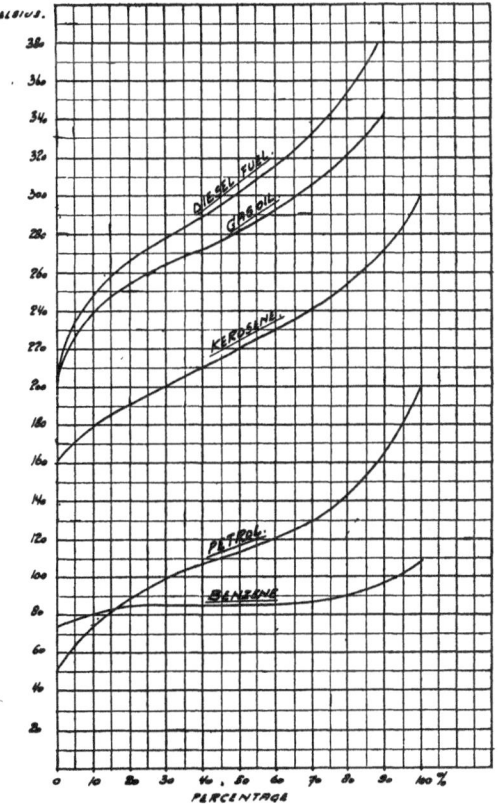

Fig. 10. Boiling ranges of fuels.

mixed with the compressed air, as a result of which complete combustion is obtained. Also when an engine is running light for a considerable time ignition will take place. With this method the compression pressure usually amounts to about 500 lbs. per sq. inch and the combustion pressure to nearly 700 lbs. per sq. inch.

B. The spray method, by which the fuel is forced into the combustion space of the cylinder direct.

The absence of the pre-combustion chamber means a simplification of the design. The energy required for atomising and mixing the fuel has to be supplied mainly by the jet of fuel itself, so that the fuel has to be injected with a higher pressure; this pressure averages about 4000 lbs. per sq. inch, whereas with the method described under A only about 1500 lbs. is needed. This higher pressure involves many difficulties, which, however, have now been overcome.

With high speed engines, especially those used under conditions where the load fluctuates a great deal, as is the case with railway engines, this spray method may perhaps not be quite so reliable.

The capacity of the pre-combustion chamber is about one-fourth to one-fifth of the total capacity of the pre-combustion chamber and the main combustion space together. When running light an engine uses only about one-fifth of the quantity of fuel used by the fully-loaded engine, so that then all the fuel may be burned in the pre-combustion chamber, where sufficient air will be available. The heat developed in this chamber and also the temperature of its walls will be about the same for the lightly and for the fully-loaded engine, as the available quantity of air in this chamber allows of only partial burning of the fuel charge. The hot wall ensures immediate ignition of the fuel in the pre-combustion chamber. Whereas with the spray method the small quantity of fuel required when the engine is running light has to be broken up in the relatively large combustion space entirely by contact with the hot air, with the pre-combustion chamber method this takes place partly through contact with hot walls. This is of much importance in cases where the load and the number of revolutions are continually varying. With heavy loads the air capacity of the pre-combustion chamber is only sufficient for the combustion of one-fourth to one-fifth of the fuel injected, the rest of it being driven by the combustion gases from the pre-combustion chamber into the main combustion space in an atomised form; the particles are so small that the engine can safely be run at a high speed without any trouble. Thus the pre-combustion chamber system ensures a good combustion also for high-speed engines at low loads.

Briefly it may be said that with the pre-combustion chamber system a reliable atomization of the fuel — which may be a source of trouble in the ordinary solid injection Diesel engines — is ensured by advanced explosion in a separate chamber. As advantages of this system may be mentioned the low pressure required for the fuel pumps, an important factor for the reliability of this sensitive part of a Diesel engine, and the insensitiveness to an inferior quality of fuel oil. On the other hand there is a drawback in that it is not so easy to start the engine as the walls of the pre-combustion chamber are then cold, and special means have to be provided for heating up.

Solid injection Diesel engines with direct fuel ignition require a high pressure (sometimes as much as 14000 lbs. per sq. inch) for injection of the fuel, and this makes the fuel pumps rather sensitive. The advantage on the other hand is that these engines can be more easily started without any special means.

Diesel engines with small cylinders sometimes give trouble when starting up cold or when run under low load. These troubles are caused by the cylinder walls being cooled off too much, in consequence of which during the compression the temperature of the air does not rise high enough to ignite the fuel at the end of the stroke.

In the Diesel engines built by the firm of Ganz & Co. at Budapest on the *Jendrassik-Ganz* system these troubles do not occur. Advantage is taken of the fact that gases of a certain pressure flowing into a reservoir in which a lower pressure exists undergo an increase of temperature. Calculations show that when, for instance, air with a temperature of 50° F. is drawn into a vacuum the temperature rises to 250 °F., which is a considerable increase.

In these *Jendrassik-Ganz* engines, which are single-acting, solid injection, four cycle, Diesel engines, at the moment of starting up the suction valves are kept closed till the suction stroke is almost completed; this special valve operation is only required until the combustion chamber is sufficiently heated, the normal valve operation being switched on after about a quarter to one-half of a minute, according to the temperature of the outside air.

Another special feature of this make of engine is in the fuel pumps, which are constructed in such a way that the discharge stroke is made under the action of released springs and only the suction stroke is effected by the movement of the corresponding cam on the camshaft. With this method of working the fuel injection is not dependent on the number of revolutions of the engine and is the same no matter whether the engine is running slow during starting up or running at normal speed. This facilitates starting up. Furthermore this allows of a wide bore in the fuel injection nozzle even for small engines; for cylinders with an inside diameter of 130 mm ($5^{1}/_{8}$") the feed nozzle has a bore of 1 mm ($\pm$ $^{3}/_{64}$").

Fuel consumption.

The fuel consumption depends on the load of the engine and increases when the engine is underloaded as well as when overloaded.

The average consumption for engines under normal load are easily remembered by the following figures:

Petrol engines . . . . 0.66 lbs. per B.H.P. per hour.
Kerosene engines . . . . 0.77 lbs. per B.H.P. per hour.
Diesel engines . . . . 0.44 lbs. per B.H.P. per hour.

Fig. 11 shows graphically how the available heat of the fuel is distributed in a Diesel engine on various loads.

Coefficient of Variation of Tractive Power.

Whereas electric locomotives have an absolutely constant torque, the tractive force of steam and motor locomotives varies according to the position of the crank. In steam locomotives the tractive force is maximum when the engine starts, for then the steam is admitted to the cylinders during the greater part of the piston stroke. The torque variation of a combustion engine depends largely on the number of impulses per revolution and on the number of revolutions per minute. With mechanical transmission the variation in tractive force will be reduced by the flywheel and a flexible clutch. When starting the locomotive with the engine running, the friction clutch compensates the torque variations.

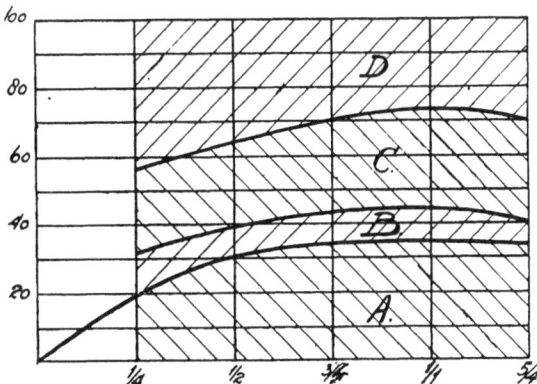

Fig. 11. Heat Balance of a Diesel engine at various Loads.

A = Effective power.
B = Mechanical losses.
C = Heat carried off by cooling water.
D = Heat in exhaust-gases.

Selection of the type of engine.

The question now is which type is to be preferred, the explosion engine or the Diesel engine. The answer is not easy, for both types have their peculiar advantages and disadvantages. The advantages of a Diesel engine as compared with an explosion engine are, setting aside the question of cost of fuel:

    a) higher efficiency (the thermal efficiency of a Diesel engine is about 30% and that of a petrol engine about 25%);

    b) the fuel is practically non-inflammable.

The disadvantages of the Diesel engine are:

    a) more bumpy running;

    b) greater weight;

    c) higher initial cost.

As a further advantage it may be mentioned that when running at a low speed the Diesel engine does not misfire so easily and develops a larger torque than the explosion engine, thus giving greater tractive force and being better able to withstand sudden resistances.

Although the Diesel engine is simpler in design than the explosion engine (no carburettor or electric ignition), technical employees are as a rule not so well acquainted with it and consequently running troubles are sometimes not so easily remedied. With trained operatives and careful maintenance, however, no trouble should be experienced. Having been used for so many years already in automobiles, the explosion engine does not present any difficulties to the personnel, so that minor defects can as a rule be immediately remedied.

Though a large number of traction vehicles are still being built with explosion engines, especially the smaller types, there is a decided tendency everywhere to adopt the Diesel engine, and as a matter of fact large-sized locomotives and motor vehicles are now being built almost exclusively with Diesel engines.

The choice is then left between two-stroke and four-stroke engines. In two-stroke Diesel engines the maximum number of revolutions is governed by the scavenging period (that is the time taken to expel the combustion gases after the working stroke), whereas in four-stroke engines the maximum number of revolutions is limited by the acceleration forces of the reciprocating parts (piston with part of driving rod); consequently a larger number of revolutions is admissible for the four-stroke than for the two-stroke engine. On the other hand, however, the two-stroke engine has smaller cylinders and if it were not for the additional weight of the scavenging pump its weight would be less. A further advantage of the two-stroke engine is that the torque is more regular, owing to the number of working strokes being twice that of the four-stroke engine.

A drawback to two-stroke engines for traction vehicles is their greater tendency to fouling.

By far the majority of Diesel engines for traction vehicles built so far are of the four-stroke type, but it is questionable whether this will be so in the future.

In order to keep their dead weight within reasonable limits Diesel engines have to be constructed as light as possible, and this can only be done by designing them as high speed engines. Up to the present, however, large Diesel engines of more than 1000 H.P. have not been built for more than 400 to 500 revolutions per minute; for lower powers up to about 150 H.P. various works turn out engines making 1000 to 1200 revolutions per minute.

With the present stage of technical development it will be possible to build for large locomotives Diesel engines with 700 to 750 revolutions per minute, without making them too heavy; the weight of a four-stroke engine may be taken roughly at 25 lbs. per H.P., but varies of course within very wide limits.

The lightest Diesel engine so far known to have been built for a locomotive weighs 2360 lbs. develops 150 H.P. and makes 1300 revolutions per minute. Beardmore's Diesel engine of 1330 H.P. (800 revs. per min) is also very light. Obviously the engine may not be made of such a light construction as would lead to defects or running troubles through insufficient strength of the various parts or inadequate bearing surfaces of axles and journals, with the risk of their running hot and being subjected to abnormal wear.

From the foregoing it will be realised that for locomotives and motor vehicles Diesel engines have to answer special requirements. Diesel engines for stationary installations are too heavy for use in locomotives and traction vehicles.

An appreciable saving in weight can be obtained by increasing the number of revolutions, the same as with electro-motors, but there is a limit also in this direction, for too high a number of revolutions leads to difficulties in securing complete combustion and keeping within admissible limits the inertial forces resulting from the moving parts.

The two-stroke system, which in itself gives a saving in weight, is seldom applied in Diesel engines for railways, because for this work low-speed engines are required, and these are necessarily heavier.

Büchi's supercharging system, in which the combustion air is forced into the cylinder with a pressure of about 5 lbs. above atmospheric pressure, increases the power by 25 to 30%.

Simplicity in operation and safe working are of great importance, for then the drivers need not be specially trained.

Further, the machinery should not make too much noise, so that the driver of a locomotive can hear sound signals and the passengers in a motor vehicle can travel in comfort; for the latter reason also the combustion should as far as possible be smokeless.

In consequence of the high mean load the coolers should be amply dimensioned.

II. TRANSMISSION.

A highly important matter in connection with traction vehicles driven by a combustion engine is the manner of transmission from the engine shaft to the driving wheels. (With steam locomotives the steam power is simply transmitted direct to the cranks of the driving wheels by means of the piston, piston rod and connecting rod, which is the best solution, for even while the locomotive is at a standstill, the full steam pressure is available on the piston so that a great starting moment can be exerted on the driving wheels. A combustion engine, however, cannot be started with its own power, but requires a separate source of energy, the power of which for starting up under load has to be relatively large, so that for this reason it is more advantageous to start the engine unloaded and then gradually increase the load. For this purpose a certain transmission is needed between the engine and the driving wheels, adjustable according to the load in so far as this cannot be arranged in the engine itself.

The following are the methods of transmission used with traction vehicles:
1. direct drive,
2. mechanical drive,
3. pneumatic drive,
4. hydraulic drive,
5. electric drive,
6. combined drive of the *Kitson-Still* system.)

DIRECT DRIVE.

In this method of transmission the connecting rods of the engine cylinders act either on a special loose crank shaft or direct on the driving cranks, and for this it must be possible to regulate the engine within wide limits, whilst it must also be reversible. Diesel engines can only be regulated by varying the fuel charge.

Though this system of transmission seems very simple, yet it has its difficulties, mainly in connection with the fact that the engine has to be built in such a way that with a normal number of revolutions the fuel charge is small, to allow of it being increased for lower speeds so as to obtain a ratio of maximum tractive force with lowest speeds. This, however, involves a larger sized engine and consequently a considerably increased weight.

Moreover the engine has to be started by means of air and in the loaded state so that a large quantity of air is required.

This system, which was applied in the first Sulzer Diesel locomotive, has now practically been abandoned entirely.

MECHANICAL DRIVE.

By this is understood the system whereby gear wheels are used in various transmission ratios, such as frequently applied, particularly for small powers. The rotatory movement of the engine shaft is transmitted via gear wheels to an intermediate shaft connected to the driving shafts of the traction vehicle by means of connecting rods or in some other way. This gear wheel transmission can be carried out in various ratios adaptable within certain limits to the variations of speed of the vehicle. The engine runs in this case with a practically constant number of revolutions, and it can be started up unloaded, by disengaging a coupling provided between engine shaft and driving gear.

For low power engines the gear transmission is arranged in such a way that the gear wheels are not continually engaged. These wheels are not brought into engagement until the engine is running, beginning with the largest ratio of reduction. When the traction vehicle has attained a certain speed the first gear wheels are disengaged and the second pair is put in with a smaller reduction ratio, and so on until the smallest ratio is reached.

For high power engines rotating gear wheels cannot be engaged so easily and the transmission is therefore designed with permanently meshing gear wheels, but the driving gear wheel is allowed to rotate loose on the driving shaft, to which it can be coupled by a special clutch. The various pairs of gear wheels are then engaged and disengaged successively. Such a driving system is applied even for very high powers — inter alia for the 4—10—2 type of Russian goods locomotive of 1200 H.P. It is not necessary that the engine should be reversible; a reversing gear can be provided in the gear transmission.

The engine is started up without load by means of compressed air or an electric starter, or by hand in the case of small engines. When the engine is making a sufficient number of revolutions the clutch can be thrown in, which should be done gradually — to avoid jumping and jerking — and should be allowed to slip a little so that the driving gear can be set in motion gradually. It is highly advisable to build in a flexible coupling.

In order to minimize wear and tear of the gear wheels they should be made of a suitable material, with the teeth accurately machined, and they must run in an oil bath.

There are generally three or four gear ratios, but for small shunting engines two often suffice for speeds of say 3 and 6 miles per hour.

Mechanical transmission can often be simplified by using chains as the last link in the transmission between engine and driving shafts, particularly in the case of small locomotives as used for shunting. In view of the shocks to which the chains are exposed it is advisable to have them made of ample dimensions and in specialised works. The life of such chains is relatively short and they have to be renewed as a rule after one or two years, but considering their small cost this is not a serious objection. Provision has to be made for taking up the elongation of the chains caused by the play arising in the hinges and the stretching of the links.

As compared with the driving system with connecting rods there is the advantage of the constant torque.

A special kind of gear wheel transmission is the **Robertson Automatic Variable Speed Gear,** as applied by Kerr Stuart & Co. Ltd. of Stoke-on-Trent, England, in their Diesel locomotives. An illustration of this is given in fig. 12. By means of a flexible coupling C the fly-wheel D of the engine drives the bush E, in which the shaft F can slide but not rotate. On the shaft F is a cam G and a metal friction disc H. There is also a sleeve J which can slide on the shaft F but not rotate on it, and to this the governor weights K K are attached, which are under the pressure of a spiral spring placed round the sleeve J. The whole rotates in an oil-tight box L. Round the sleeve J there is also a collar M supported by a ball-bearing. By moving the handle N, the collar M and the sleeve J can be shifted along the shaft F and fixed in a certain position. There are two friction rings O and P at the head end of the gear box Q, which can oscillate around the shaft R. This gear box Q contains the gear wheels SSS, by means of which the rotary movement of the disc P is transmitted to the shaft R, around which the gear box can oscillate freely in a vertical plane against the plane of the friction disc H. The friction ring O can move floatingly between the friction plates H and P. The necessary friction pressure between the friction disc H, the ring O and the disc P is obtained from the centrifugal force of the governor weights KK.

Fig. 12a. Robertson automatic variable speed gear. Section.

The action is as follows: when the engine is running light and the locomotive is at a standstill the centrifugal force of the weights KK is not sufficient to overcome the tension of the spring, so that there is nothing to prevent the cam G from being drawn to the left, thus disengaging the friction disc H from the ring O. By throwing over the lever N the position of the sleeve J and, with that, of the weights KK can be changed. In the no-load position when the locomotive is stationary the lever N draws the collar M right over to the left, and with that the friction disc H, so that there is no pressure between the friction plates. The lever N can be set in various positions.

The locomotive is set in motion by pulling the handle N to the left, which sends the collar M to the right and the friction disc H is pressed against the ring O. As long as the engine is running slowly the weights KK do not come into play and consequently the friction wheel H does not bear on the ring O, but as soon as the engine gets up speed the weights KK come into action and a pressure is set up between the friction disc H and the ring O, the former trying to rotate the latter. The inertia of the train, however, will stop the shaft R from turning and the gear wheels SSS will swing around this shaft, carrying the box Q with them, with the result that the contact between the ring O and the disc H is reduced to a smaller diameter of the latter. This changes the transmission and when a sufficiently small diameter of H is reached the inertia of the train is overcome and the locomotive is set in motion. But as soon as the train begins to move the reaction moment which caused the box Q to rotate around the shaft R is reduced and owing to the frictions (assisted by the higher speed of the engine and the wider throw of the weights KK) the ring O begins to work up towards the largest diameter of the wheel H, this increasing the number of revolutions of the shaft R and consequently the speed of

the locomotive, until eventually the ring O reaches a position concentrical with the disc H, when the coupling acts as a simple friction plate-coupling.

In case a gradient has to be overcome which is too steep for the speed at which the locomotive is running, the increased reaction moment will cause the box Q (via gear wheels SSS) to swing around the shaft R, and through the reduction in the number of revolutions of the engine resulting from the greater resistance the throw of the weights KK will decrease and the transmission ratio is again changed by the ring O acting again on a smaller diameter of the disc H until the reduction of the speed of the engine corresponds to the work to be done.

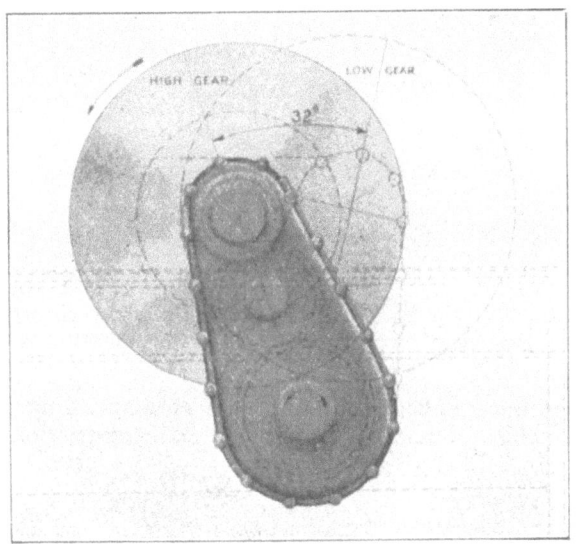

Fig. 12b. End view.

As may be seen from the illustration, the ring O runs quite free on disc H and its rotatory movement is transmitted to the disc P. The object of this is to allow the ring O always to rotate independently of the power that has to be transmitted. Any slip will take place on the ring P, and for this reason the friction coefficient of the ring P is smaller than that of the ring O, thus avoiding the grinding of grooves in the friction disc O. By changing the relative position of the weights KK by means of the cam G, the rate at which the transmission automatically adjusts itself is likewise changed, and this can be done by moving the handle N. The forward and reverse movement is governed separately by engaging a left hand or a right hand set of bevel gear wheels.

This Robertson automatic speed gear is made for powers up to 30 H.P.

In the **SLM Winterthur transmission** invented by H. Jenny (see "Schweizerische Techniker Zeitung", vol. XXII, No. 20) all the gear wheels are continually engaged. The various ratios of transmission are put in and thrown out by means of friction discs built-in in the gear wheels and operated under oil pressure.

The principle of the construction and method of operation are illustrated in fig. 13. Pinions I, II and III for the reverse movement and for the first, second and third speeds of the forward movement are keyed into the shaft *b* and these are in

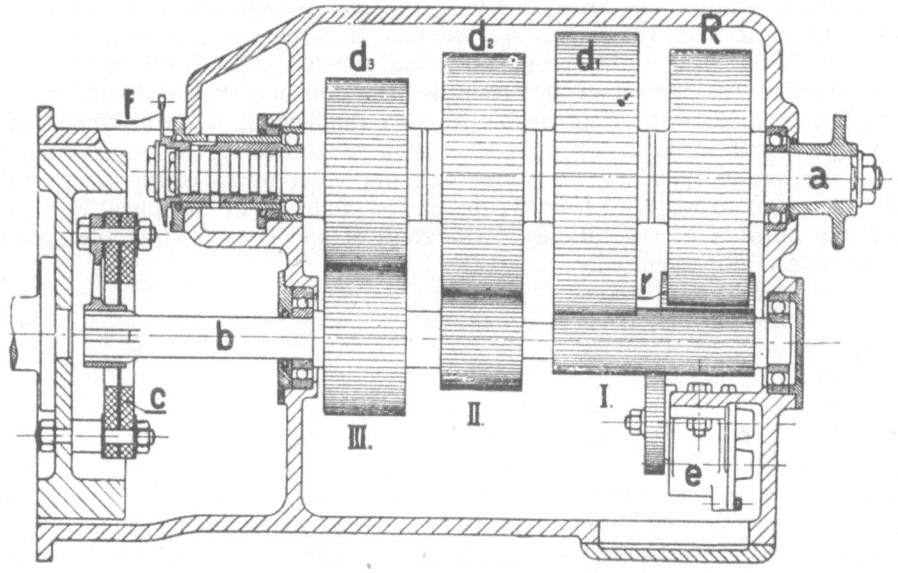

Fig. 13a. Oil pressure change-speed gearing.
Diagram of an oil pressure change-speed gearing for 3 forward speeds and 1 reverse speed.

continual engagement with the corresponding gear wheels R, d_1, d_2 and d_3, each of which is made in two halves with the friction couplings in between (see figs. 13b and 13c).

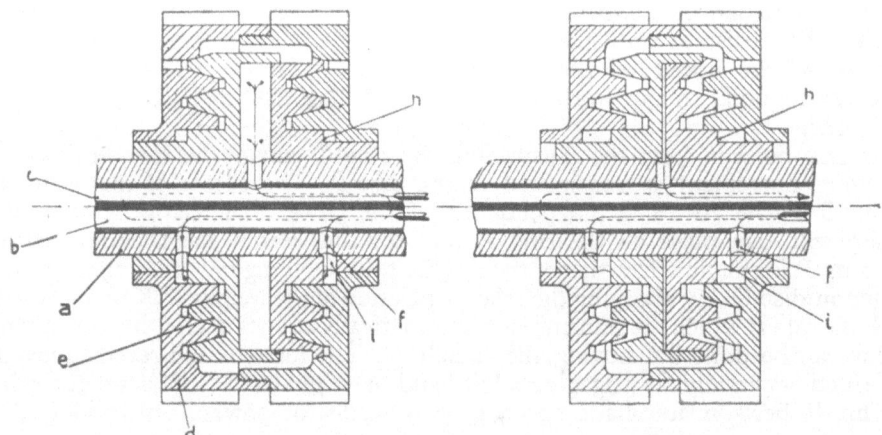

Fig. 13b. Coupling engaged.　　　　Fig. 13c. Coupling disengaged.

The oil pump *e*, which is driven by pinion I, draws the oil from the bottom of the gear box and forces it under pressure (the pressure may be 30 to 85 lbs. as required) to the main distributing valve *f*, by means of which the flow is regulated to the gear wheels. When the engine is running light the valve cuts off the oil supply to the gear wheels. The oil subsequently flows back into the gear box and from there is circulated again by the pump.

The working of the various couplings may be seen in figs. 13b and 13c. The double gear wheel *d* runs loose on the boss of the two friction discs *e*, which are grooved so as to drive the coupling shaft *a*, on which they slide axially. For engaging the coupling the pressure oil is led through the channel *c* on the shaft *a* between the coupling discs, these then being forced up against the rotating gear wheels and gradually turned with them. In the disengagement of the gear the oil is drawn out of the space between the friction discs through channel *c*, and in consequence of the pressure in the chambers *h* the discs slide together. There is a continuous supply of pressure oil to these chambers *h* via the channel *b* and the bores *f* and *i*. When one speed is put in the previous one is automatically thrown out at the same time, so that two gears can never be engaged simultaneously. For motor locomotives and vehicles the gear wheel transmission may be coupled to the engine direct, in which case the driving shaft is connected to the

Fig. 14. S.L.M. Gearing enclosed.

gear by means of a link, while a special box is provided. It is also possible for the whole of the gear to be combined in one single box supported by the driving shaft, in which case the gear wheel box is suspended from one or two other points in the frame to allow of it following changes in the form of the frame. As a rule the connection with the engine is made by means of a flexible coupling.

The gear box is made of aluminium, cast steel or cast iron; it must be absolutely oiltight. The shafts, all of which are mounted on ball bearings, the gear wheels and the bearings are always in an oil bath. The pressure of the oil from the pump can be regulated between zero and the maximum by means of a special valve, independently of the number of revolutions of the engine. In this way shocks can be avoided when the driver changes speeds; when starting, for instance, he can reduce the oil pressure and increase it gradually, so that the friction discs are pressed against each other with a uniform increase of force.

Figs. 14 and 15 show this kind of transmission as applied to a motor vehicle of 100 H.P. The part marked *a* is the pinion shaft, the end of which is made in the form of a universal joint with connecting flange. To this the engine is connected by means of a countershaft. At the end of the control-shaft *b* is the distributing valve *c*. This control-shaft transmits the movement to the reversing gear. A counter-shaft *d* is fitted with two large bevel gear wheels *e* which are rotatable on the shaft and in mesh with a conical pinion *f* on the end of the control shaft. Thus the two big gear wheels always rotate in opposite direction. On the inside these bevel gear wheels are made in the form of a dog-clutch. On the same counter shaft a small gear wheel is fitted, which, by means of a fork and a lever, the driver can slide along longitudinal grooves in an axial direction; this is likewise provided with claws on both sides and meshes with the large driving wheel *g* on the driving axle. Consequently the direction of rotation of the driving axle and thus the direction of travel of the vehicle will always be variable according to which of the large bevel wheels is engaged by the small wheel, whilst the control shaft is always rotating in the same direction.

Fig. 15. S.L.M. Gearing open.

Important advantages of this oil pressure transmission are the possibility of being able to reverse direct and the braking by means of the driving gear. Seeing that the various gear wheels are continually in mesh and in the changing over from one speed to another the only coupling mechanism consists of metal discs rubbing against one another, there are no shocks in the gear, not even when the direction of travel is suddenly reversed. Through the rubbing of the friction surfaces against each other, when reversing, the inertia of the vehicle is transformed into frictional force, so that prior to the actual reversal a strong braking action is set up in the driving gear itself. In this there lies a certain advantage from the point of view of safety: in case of an imminent collision, for instance, first the brakes have to be applied, and in this case with a vehicle fitted with an ordinary mechanical drive the driver has to disengage the coupling between the engine and the gear, throttle the fuel supply, apply the brakes and maybe reverse the gear all practically simultaneously, whereas with the transmission described above all the driver has to do is to throw over the reverse handle and the braking begins at once. The braking action is as follows: at the moment of reversing, owing to its inertia, the vehicle continues in the same direction and carries the control shaft with it; on the reversing handle being thrown over the friction discs for the backward movement are pressed up against the gear wheels, which are running in an opposite direction, thus setting up friction between the two; the inertia of the car is converted into heat, which is carried off by the circulating oil, so that the driving gear is not excessively heated.

No shocks occur when reversing. The efficiency of this transmission is about 95%, and it can now be made for powers up to 300 H.P.

The **Soden gear wheel transmission** is arranged for five speeds. In the gear box (see fig. 16) there are three shafts, A, C_1 and C_2, on ball or roller bearings. The engine drives shaft A and a shaft B in line with A transmits the motion to the driving wheels. On each of the shafts C_1 and C_2 there are two loose gear wheels (b and i and f and l respectively) and one fixed gear wheel (c and g respectively), whilst shaft A has fixed gear wheels a, e, k and h. The respective sets of gear wheels are continuously in mesh, but the loose wheels can slide along the shafts C_1 and C_2, with which they can be engaged by means of claw couplings K_1 and K_2. The operations for changing speeds are as follows:

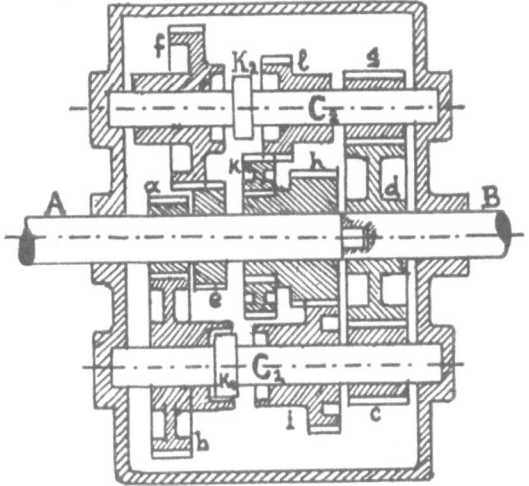

Fig. 16. Soden gear.

Speed	Gear wheels to be moved	Transmitted by gear wheels	Ratio of transmission
I	b to the right	$a : b \times c : d$	1 : 7.11
II	f ,, ,, ,,	$e : f \times g : d$	1 : 4.37
III	i ,, ,, left	$h : l \times g : d$	1 : 2.66
IV	l ,, ,, ,,	$k : i \times c : d$	1 : 1.63
V	shaft A coupled direct to shaft B		1 : 1

Each of the gear wheels is shifted by means of a fork G on a sleeve H, which is shifted by a shaft D (see fig. 17). At the top end of the fork G is a pin S, which enters a hole in a rotatable roller W when this is in a certain position. A spring F keeps the fork G to the right, in which position the gear wheel on the shaft C is coupled with the shaft by means of the claw coupling K. In the position shown in fig. 17 the claw coupling is not engaged, owing to the fork G being pressed to the left by the lever on the shaft R, which is brought into this position by the pressure of the air in the cylinder Z, against the action of spring F.

There are five of such coupling arrangements side by side, one for each speed, and the roller W has likewise five holes, so that only one of the pins S can be moved to the right at a time, thus avoiding two speeds being put in at the same time.

The shaft A (see fig. 16) is connected to the engine by means of a multiple disc coupling. When this coupling is free, the forks G are shifted to the left by the shafts R and consequently all transmissions are thrown out. Thus the roller W is free

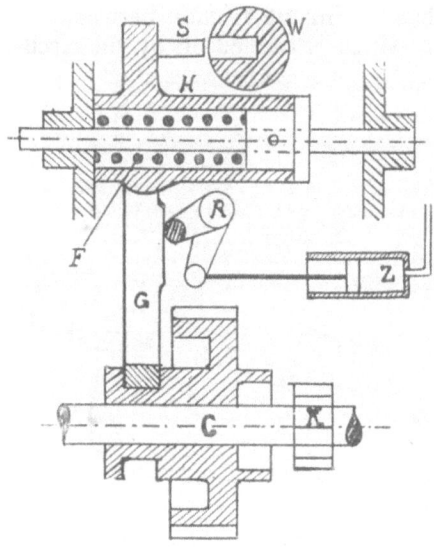

Fig. 17. Soden gear.

and can be switched in for a certain speed. The arrangement is such that when a certain set of gear is engaged the multiple disc coupling is closed and the motion of the engine is transmitted to shaft B (fig. 17). In the changing of gear in the Soden transmission several actions have to take place in quick succession.

1) The main (multiple disc) coupling is closed by the force of springs and released by air pressure.
2) Mechanically connected to the main coupling is the cut-out R, which forces back or releases the forks of the gear wheels to be shifted, so that the spring can come into play.
3) It depends on the position of the roller W which speed gear is put in.
4) When only one engine has to be attended to by the driver the roller W is rotated by direct mechanical means, but when several engines have to be operated by one driver it is rotated by air pressure or electricity.

Transmission made by the "Deutsche Getriebe G.m. b.H." of Hannover.

Power units for motor coaches have been built with Diesel or petrol engines up to 150 H.P. and sometimes two such power units are used together, thus giving an available power of up to 300 H.P.

The transmission from the engine to the driving axles consists of gearing and cardan shafts, different gears being used according to the engine power, weight of the vehicle, number of driven axles, etc.

The engine with the gearing and the necessary auxiliary apparatus is built in a special frame suspended from the frame of the motor coach by means of springs in a noise-absorbing manner. This saves space in the motor coach itself, running shocks are not transmitted to the engine and the machinery is easily accessible.

There is a direct drive from the engine to a dynamo for the supply of electric current for lighting via an accumulator, and if necessary to a compressor for the supply of compressed air for the brakes and other pneumatic devices. The engine is started electrically or by compressed air according to the type of engine used; small engines are also provided with a hand-starter.

The coupling between engine and gear is a dry multiple disc coupling.

There are four speed variations with ratios of about 1 : 5, 1 : 2.8, 1 : 1.7 and 1 : 1 for both directions. All gear wheels are continuously in mesh. In the large types of gear the reversing gear is combined with the speed changing gear, the former being placed in front of the latter, so that the faces of the gear wheel teeth are used alternatively in the two directions, thus equalising the wear on both faces and about doubling the life of the gearing.

The shafts and wheels run in an oil bath in a dust-proof box (fig. 18).

The changing of speeds is arranged on the *Mylius* system with combined friction and claw couplings. With this system the driving force is transmitted by claw couplings, which are first automatically synchronised with the aid of small friction couplings and consequently are engaged shock-free. The gears are changed quickly and a smooth action is ensured independently of the skill and care of the driver; also when changing back to lower speed the claw couplings work without shocks

Fig. 18. Reversing and change-speed gearing, system Mylius, for oil engine vehicles.

or noise. The auxiliary couplings for synchronising the claw couplings serve merely to accelerate or retard the free-running gear parts to the speed required, not transmitting any of the driving force of the engine to the vehicle. Consequently these couplings are subject to hardly any wear at all, so that the gearing requires little maintenance.

Fig. 19. Shaft drive with reversing gear.

In the small types the gear is changed mechanically and in the large types pneumatically or electrically, so that several motor coaches can be coupled together and directed from one driver's cabin by means of train couplings.

The axle drive consists of a bevel wheel gearing or in special cases a combined bevel and spur wheel gearing.

In the types of gearing without reversing gear the axle gear is designed as reversing gear, one or the other of two wheels provided on the driving axle, according to the direction, being coupled to the axle by means of a dog clutch. The bevel wheels of this reversing gear are permanently engaged. To reverse all that is required is to slide the claw coupling over the shaft (fig. 19).

The arrangement of the machinery differs for various types of motor coaches:
a) For light motor coaches and rail motorbuses (fig. 20).

The power units may consist of Diesel engines up to about 80 H.P. or petrol engines up to 150 H.P., being destined for light cars and especially for nar-

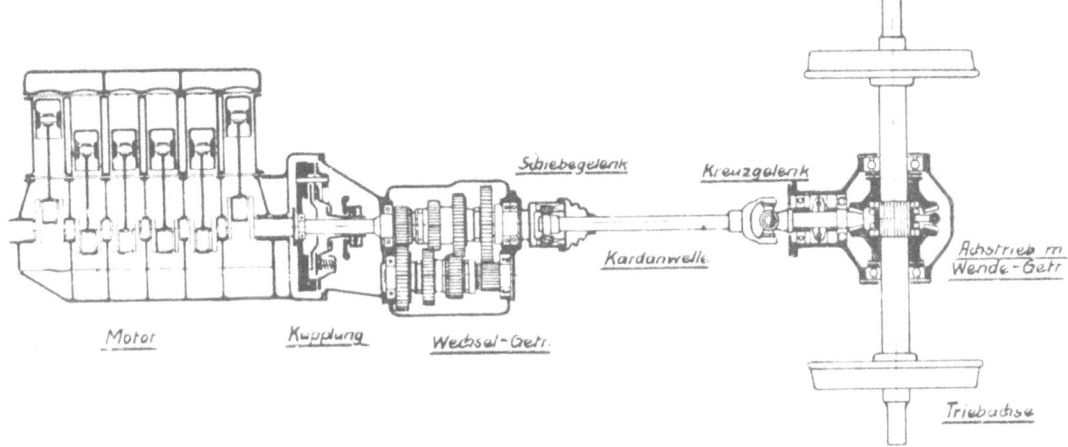

Fig. 20. Power unit.

Achstrieb m. Wende Getr. = Axle drive with reverse gear. — Kardanwelle = Cardan shaft. — Kreuzgelenk = Universal joint. — Kupplung = Coupling. — Motor = Engine. — Schiebegelenk = Slide joint. — Triebachse = Driving axle. — Wechselgetriebe = Change-speed gear.

row gauge. The vehicles can, however, also be fitted with two engines to reach a capacity of up to 300 H.P.

The general arrangement is similar to that described above. The reversing gear is embodied in the axle-drive; the control of the engines takes place mechanically by means of rods from the two driver's cabins (at each end of the car); in special cases, for instance when two engines are provided, or when for other reasons the driver's work has to be simplified, pneumatic operation is provided.

Each engine drives one axle, but for cars with bogies, when a larger adhesion weight is needed, also the second axle is driven by means of chains or cranks; thus the drive is on half the number of axles of a car with one engine and on all of them in the case of a car with two engines.

A special design with auxiliary intermediate transmission allows of an all-axle drive also for single engine cars.
b) For heavy vehicles doing heavy duty (fig. 21).

As power units petrol or Diesel engines up to 150 H.P. at 1200 revolutions per minute may be used, the entire machinery being specially designed to meet heavy demands. One or two engines are used according to the tractive power and speed required. The double-engined unit is recommendable because it gives greater security against breakdowns, for then the motor coach can be kept running even if one of the engines should break down.

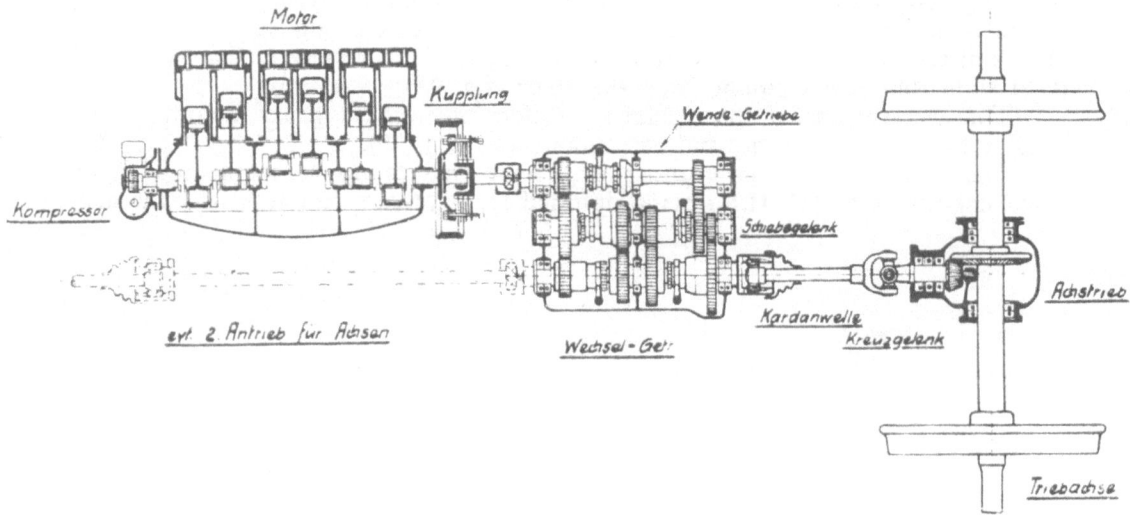

Fig. 21. Power unit.
Achstrieb = Axle drive. — Evt. 2. Antrieb für Achsen = 2nd axle drive (if required). — Kardanwelle = Cardan shaft. — Kompressor = Compressor. — Kreuzgelenk = Universal joint. — Kupplung = coupling. — Motor = Engine. — Schiebegelenk = Slide joint. — Triebachse = Driving axle.

Mylius system of change speed gear for rail motor-buses (fig. 22).

With this system there are four different speeds and all gear wheels are permanently in mesh.

The gears coupled for the various speeds are as follows:

1st speed: gears A—B and C—D by engaging wheel D in the claws of the main shaft K.

2nd speed: gears A—B and E—F by engaging wheel E in the claws of the main shaft.

3rd speed: gears A—B and G—H by engaging wheel C in the claws of the counter shaft.

4th speed: gear wheel A direct with the main shaft K by engaging sleeve J in the claws of wheel H.

Gear A is made in one piece with the driving shaft driven by the main coupling. Gears B, C, H and F are keyed on tight to their shafts, whereas gears D, E and G are loose and movable.

The coupling is done by means of the four striker rods 1—4, the forks of which engage in the grooves of the movable gear wheels D, E and G and of the sleeve J, which latter has claws on either side. All these movable gear wheels and the sleeve J carry on the left-hand side the cone of a friction coupling and on the right-hand side a toothed claw.

When changing speed the gear required is moved to the left and by means of the friction cone becomes engaged with the part of the shaft carrying the counter

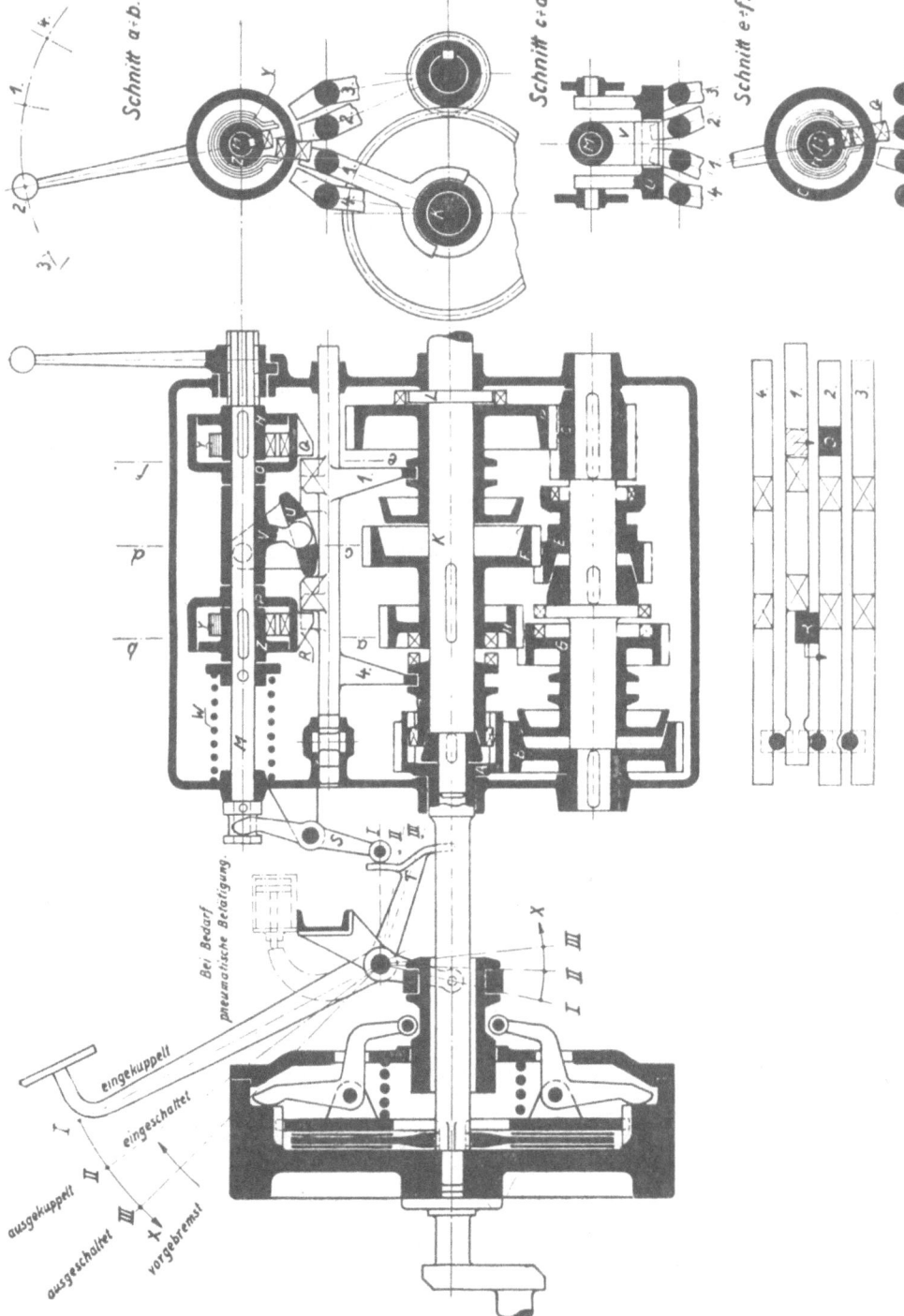

Fig. 22. Change-speed, gear System Mylius, Diagram.

Ausgekuppelt = Coupling disengaged. — Ausgeschaltet = Coupling thrown out. — Bei Bedarf pneumatische Betätigung = Pneumatic operation if required. — Eingekuppelt = Coupling thrown in. — Schnitt = Section. — Eingeschaltet = Coupling engaged. — Eingekuppelt = Coupling thrown in. — Vorgebremst = Advanced braking.

claw; this equalizes the speeds of the two claw halves (preliminary braking). On the next succeeding movement to the right these claw halves now rotating at the same speed are coupled up without shocks and the speed is changed.

This reciprocating movement of the striking rods answers to the movement of the clutch pedal, so that on the clutch being disengaged the preliminary braking is effected and on the clutch being engaged the claws are coupled together.

The action of the transmission is as follows:

Each striking rod has two catches which are engaged by the fingers R and Q of the boxes P and O; these boxes are rotatable but are fixed on the control-shaft in such a way that they cannot slide over it in an axial direction. On the pedal being pressed down the control-shaft is moved to the left by the lever S and the curvature T on the pedal-lever, and on the pedal being released it is moved to the right by the spring W.

The sketch shows the first speed gear engaged by the coupling of gear wheel D with claw L. If the speed is to be changed from first to second the change speed lever is put into second (control by notched quadrant) and through the resultant rotation of shaft M the box O is likewise rotated (by means of the flanged hub N and the spring Y) and the finger Q is moved from striking rod 1 to striking rod 2 (see section e—f and plan). The finger R cannot follow this movement on account of the box P being prevented from rotating by the catch of striking rod 2. Consequently the hub Z stretches only the corresponding box spring Y (see plan and section a—b).

Speed is changed by the releasing and engaging of the coupling in the following manner:

A. Release of coupling.

Pedal moved from I to II.
1. Disengagement of coupling.
Pedal moved from II to III.
1. Idling of the coupling sleeve.
2. Disengagement of first speed by shifting control-shaft M and by the centering device V—U (its function is explained later).

In this centre position the stretched spring Y pulls the box P with finger R onto striking rod 2.

Pedal moved from III to X.
1. Idling of the coupling sleeve.
2. Finger Q engages the striking rod and presses in the friction coupling of the second speed (preliminary braking).

B. Re-engagement of clutch.

Pedal moved from X to III.
1. Idling of the coupling sleeve.
2. Disengagement of the preliminary brake coupling of the second speed by spring W.
Pedal moved from III to II.
1. Disengagement of coupling sleeve.
2. Engagement of second speed by spring W.
Pedal moved from II to I.
1. Engagement of coupling.
2. Control-shaft M becomes stationary.

Any desired change of speed is obtained in the same way, the perfect equalisation of speeds by preliminary braking ensuring a proper coupling when changing over both to higher and to lower speeds. Of course this process of changing over takes place very quickly and immediately responds to a vigorous thrust and quick release of the clutch pedal.

The centering device V—U (see longitudinal section and cross section c—d) consists of a finger V which can rotate around the control-shaft M and engages in a slit of the rocking swing U, which it causes to swing out in the same direction in which the shaft M is displaced. When V swings back the non-centred striking rod is pushed back at its catch until centred, so that the swing U can oscillate over the catches of the rods.

The striking rods are locked against each other, so that only one rod can be in action at a time.

The various positions of the speed variation lever are fixed by notches for each speed.

Speeds may, of course, also be changed by means of push buttons, rotary controllers or similar devices, and the coupling and control may also be easily operated by air pressure or some other servo-action as desired, there being no need of a fine touch for changing speeds.

Pneumatic control gear for oil-engine vehicles with the Mylius gearing system (fig. 23).

With this pneumatic control several motor coaches coupled together or a motor coach with trailers can be controlled from any driving position (distance control).

The speed is changed and the coupling controlled by means of an air-pressure valve (control valve) in the driver's cabin. The pipe connections 1 to 4 lead to small air cylinders with which the locking bolts W and S for any of the four striking rods J can be lifted. Connection 5 leads to the coupling cylinder L, the inlet and outlet of air being controlled indirectly by corresponding valves.

For each of the four speeds 1 to 4 and for the zero speed the control valve has two positions, viz.:

1. the preliminary position, in which the locking bolts W and S are lifted while the coupling is disengaged;
2. the running position, in which the locking bolts are likewise lifted while the coupling is engaged.

In the zero position all bolts are free, so that no speeds can be changed or engaged; this position serves for placing the gear in "neutral".

The striking rods are moved by air pressure from the cylinder L and by the spring C.

When the air is drawn out of cylinder L the spring C cuts out the engine coupling K and the speed gear last used. If at the same time the clutch is thrown in for a certain speed the spring C presses in the friction coupling a of that speed gear and synchronises it for engagement (preliminary braking).

On air being admitted into the cylinder L the speed gear required and thereupon the engine coupling are engaged.

The changing over is effected in the following manner: the position shown in the figure 23 corresponds to the fourth speed, the crank of the control valve being in position 4 (the clutch is thrown in and the locking bolts W_4 and S_4 are under pressure). Thus the coupling claw z for the fourth speed and the engine coupling K are engaged.

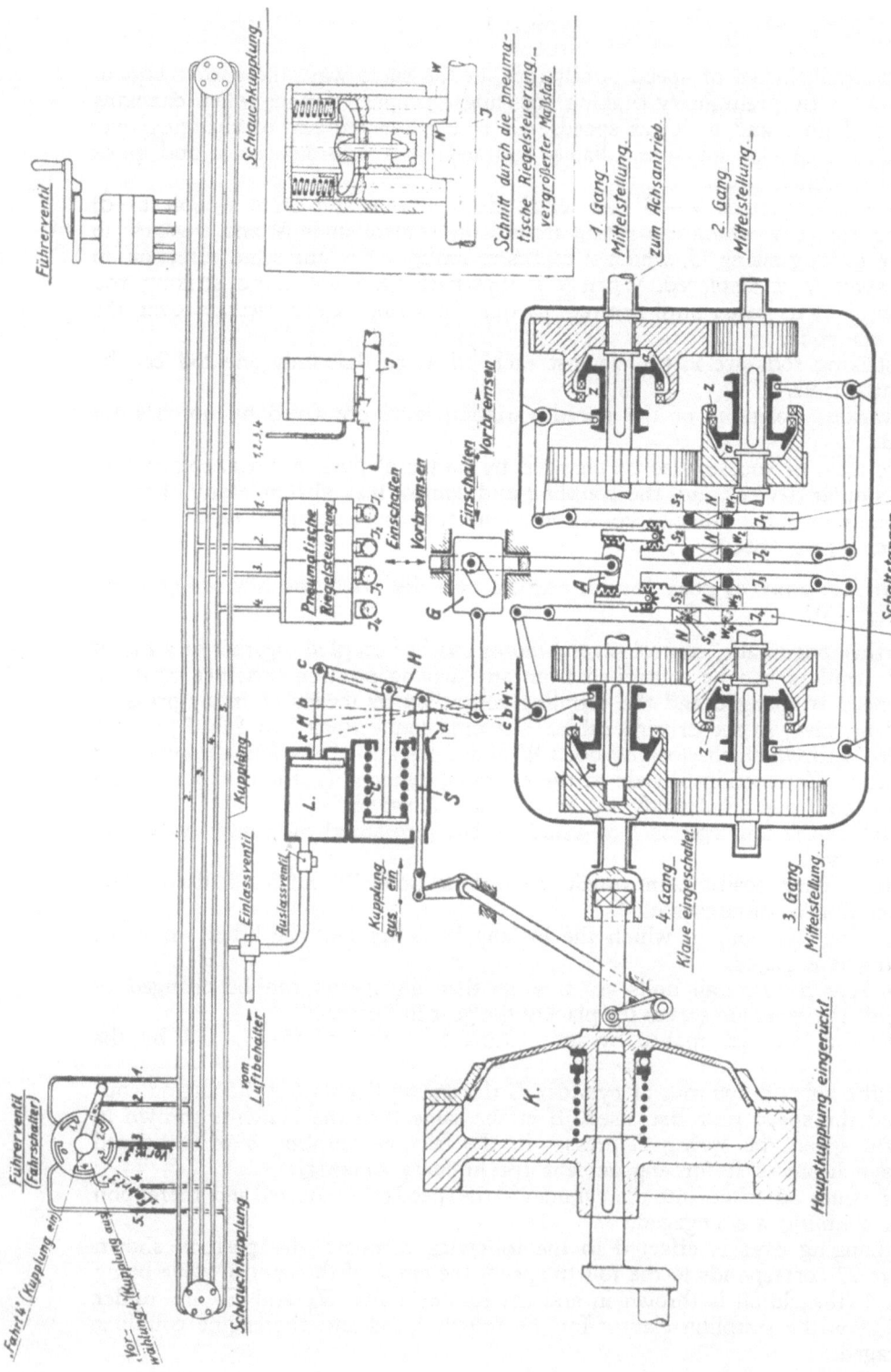

Fig. 23. Pneumatic control gear for oil-engine trains, System Mylius.

Kupplung aus = Coupling disengaged. — Kupplung ein = Coupling engaged. — Auslassventil = Exhaust valve. — Einlassventil = Inlet valve. — Einschalten = Coupling. — Fahrschalter = Controller. — Fahrt 4 (Kupplung ein) = 4th speed (coupling engaged). — Führerventil = Control valve, master valve. — Gang = Speed. — Hauptkupplung eingerückt = Main coupling engaged. — Klaue eingeschaltet = Claws engaged. — Kupplung = Coupling. — Mittelstellung = Centre position. — Pneumatische Riegelsteuerung = Pneumatic lock control. — Schaltstangen = Striking rods. — Schlauchkupplung = Hose coupling. — Schnitt durch die pneumatische Riegelsteuerung = Section through pneumatic lock control. — Vorbremsen = Preliminary braking. — Vergrösserter Masstab = Enlarged scale. — Vom Luftbehälter = From air vessel. — Vorwählung 4 (Kupplung aus) = Preliminary choice of speed 4 (coupling disengaged). — Zum Achsantrieb = To the axle drive.

Suppose, for instance, that the gradient is too steep for fourth speed, so that the driver has to change over to third. The crank is then quickly thrown over to advance position 3 (coupling released, bolts W_3 and S_3 under pressure, bolts W_4 and S_4 free). Bolts W_4 and S_4 descend, while bolt W_3 is lifted. At the same time the air is drawn out of the cylinder L, the spring C pulls lever H from position c—c to the left into position b—b and thus disengages engine coupling K by means of rod S.

As soon as the engine coupling is disengaged rod S is stopped by the fixed point d, forming a fixed fulcrum for lever H, whilst the spring C then forces the lever further into the positions M—M and x—x. This movement of lever H sends the slide G to the right and forces down the striking rods J_1 to J_4. Striking rod J_4 moves downward into the centre position M, thereby disengaging the toothed claw of speed 4, and is held there by bolt W_4, which in the meantime has descended. Striking rod J_3 moves downward underneath the lifted bolt W_3 into the preliminary braking position x and presses in friction coupling A of the third speed. Rods J_2 and J_1 are held in the centre position by bolts W_3 and W_1. The change of pressure releases bolt S_3, which is then raised and leaves the passage free for striking rod J_3.

After the air has escaped from the cylinder the drive throws over the crank into running position 3 (bolts W_3 and S_3 under pressure and coupling engaged). Bolts W_3 and W_4 remain in the raised position, air is admitted into cylinder L, spring C is compressed, rod S forms the fixed fulcrum for the lever H and all striking rods are raised by means of the slide G.

Striking rods J_4, J_2 and J_1 are then held in position by bolts S_4, S_2 and S_1, while rod J_3 is moved upward, releases the friction coupling of the third speed and engages the claw coupling in position b—b. This is the outermost position of the piston G and thus the bottom end of lever M becomes the fulcrum. The air pressure raises the rod S from its fixed point d and on its way from b—b to c—c engages the engine coupling again, thus completing the process for the change of gear.

In the same way the speed can be changed into any other speed by first throwing over the crank into the advance position and then, after the air has escaped, into the running position for the desired speed.

While the vehicle is running the crank is placed in Running Position O, all bolts W and S then being released from air pressure, while coupling K remains engaged. The engine is stopped by turning the crank to Advance Position O, which disengages coupling K and places the gear in the zero position.

PNEUMATIC TRANSMISSION.

With this type of transmission the combustion engine drives a compressor which supplies compressed gas to cylinders arranged similarly to those in steam locomotives, thus giving the same flexible working as obtained with steam engines.

The "Maschinenbauanstalt Görlitz" built, by way of experiment, a 100 H.P. Diesel locomotive in which the exhaust gas of the engine was used as the means of transmission. The exhaust gas was compressed and subsequently expanded in the cylinders of the locomotive, then compressed again and so on, thus being continuously recycled. The exhaust gas was chosen as power transmitter in order to avoid the risk of the lubricating oil exploding at the high temperatures arising in this system, there being no oxygen in the exhaust gas. While the locomotive was stationary but the Diesel engine running the gas passed through a by-pass of the suction and discharge valves, this by-pass being closed more or less when the loco-motive was running. With this system starting takes place very gradually.

The efficiency of this trial locomotive, however, was not very high, only about $9\frac{1}{2}$ to 10%. Moreover the exhaust gas contained too many impurities, which was bad for the valves. In order to increase the efficiency it is necessary that the temperature of the gas used as transmission agent should be as high as possible when the gas enters the locomotive cylinders and as low as possible when leaving the cylinders.

It was therefore evident that it is better to use atmospheric air as power transmission agent, especially since it has been found that with a proper design there need be no danger of the lubricating oil exploding.

For heating the compressed air it is quite possible to utilise to advantage the heat of the exhaust gas of the Diesel engine, which contains about one-third of the heat available in the fuel. The air then supplied to the locomotive cylinders does more work than is transmitted by the shaft of the Diesel engine.

According to *Zarlatti*'s system atmospheric air is compressed to 115—170 lbs. per sq. inch and mixed with water heated by the combustion gases from the engine. The water evaporates, leaving the air saturated with steam. The air is compressed in several stages in order to avoid too high a temperature in the compressor.

HYDRAULIC TRANSMISSION.

By hydraulic transmission is understood the transmitting of motion from the engine shaft to the driving wheels by means of a liquid (oil). There are various types of this system, but the typical feature of all of them is the driving of one or more oil pumps by the combustion engine. The oil under pressure supplied by these pumps (the active pumps) provides the driving power for one or more hydraulic engines (passive pumps) coupled to the driving wheels. The advantage of this method of transmission lies in the smooth drive, whilst the installation of the active and passive pumps is very simple, being connected merely by pipes. For the transmission of high powers, however, very large pumps are required, thus involving a considerable increase in weight.

Fig. 24a. Lentz gearing. Elevation.

Lentz Gearing.

The design and action of this gearing (see fig. 24) is as follows: the speed of the engine shaft W is transmitted to the driving axles of the traction vehicle by means of an oil pump P and an oil engine M. The pump P is connected to the engine shaft W, and the engine M to the driving axles by means of the coupling rod T or in any other way. The transmission of the two pumps and consequently the speed of the vehicle can be varied by regulating the volume of liquid passing from the pump P to the engine M, this being effected by means of the regulating valve St. With the aid of the reversing valve V the direction of rotation of the shaft of the engine M can also be changed for forward or backward running; the engine shaft W and the shaft of the first pump always rotate in the same direction.

Both the pump P and the engine M consist of a cylinder in which a heavy concentrically mounted shaft rotates. The shaft is provided with slots into which blades S can slide radially in such a manner as to normally fit up tightly against the inner wall of the cylinder. These blades are drawn back when passing from the discharge to the suction side, thus bringing about the desired pump action; otherwise there would be no displacement of liquid from the suction to the discharge side or vice versa and there would only be a circular flow of liquid in the pump.

Pump P has several chambers, for instance four, one behind the other. By adjusting the valve St, the oil from one or more of these chambers can be either delivered to the engine M or returned to the suction chamber of pump P.

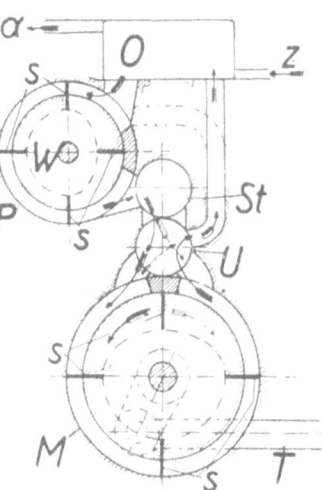

Fig. 24b. Lentz gearing. Diagram.

For the lowest speed the oil of only one of the chambers in pump P is delivered to the engine M, the oil from the other chambers being led back to the suction side of pump P. A small volume of liquid is then delivered to the engine M, but under high pressure; the engine M thus runs slowly but produces a considerable torque. When all the chambers of pump P are connected up then the largest volume of oil is delivered to the engine M, but under less pressure; the engine M then attains its maximum speed, but the torque is a minimum. Intermediate speeds and torques are obtained by connecting up an intermediate number of chambers. In this manner the transmission ratio is varied step by step. Changing over, however, can be done under load. It is not necessary to build in the pump P and the engine M in one box; they can be mounted separately and connected up by pipes.

Practical experience has shown that the efficiency at low speed (high oil pressure) is 65—70% and at high speed (lower oil pressure) 75—80%.

Schwartzkopff-Huwiler Gearing.

This device is shown in fig. 25. The pump consists of a rotor 1 with slots in which the blades 2 can move radially. The blades have pins with rollers on each side which are guided in grooves 3 in the side covers. A cylinder 4 (regulating cylinder) with grooves for passage of the blades can slide over the rotor; this cylinder rotates together with the rotor and with this the size of the working chambers of the pump (the space between the blades when these are standing out) can be varied from nil to the maximum. When the cylinder is pushed completely into the working space it closes it entirely and no oil is then delivered by the pump, and the engine is at a standstill. When the cylinder is drawn out to its fullest extent the working chambers are open to their maximum capacity and the engine runs at its highest speed. Between these two extreme positions of the cylinder 4 any desired position can be taken, thus allowing of any variation of the speed of the engine.

The cylinder is adjusted by means of the screw spindle 7 or in any other way. When the cylinder moves in and out it moves also the cut-offs 6 between the suction and delivery chambers.

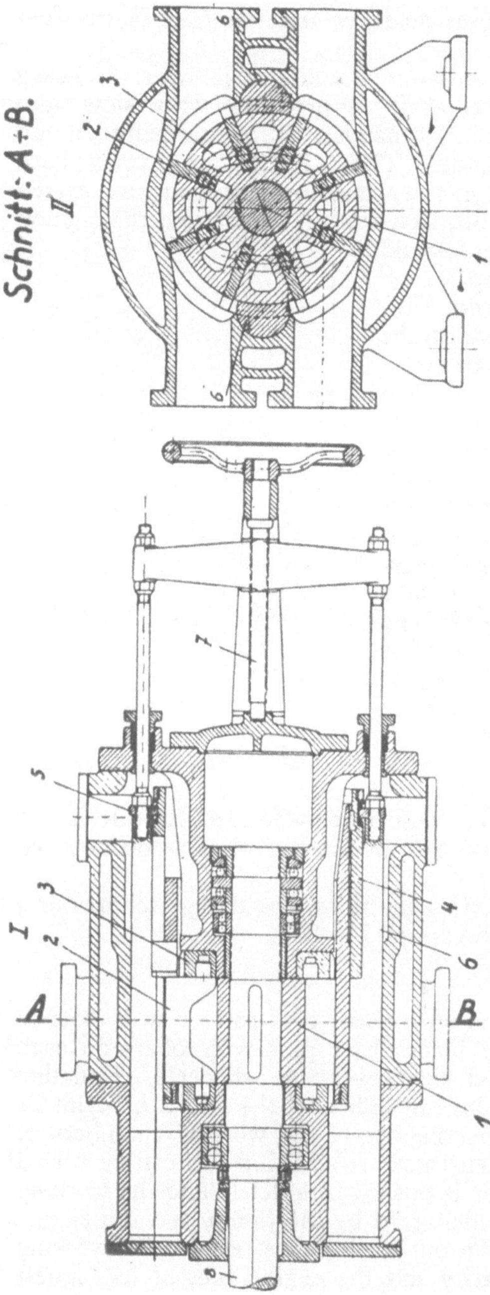

The pump is always made double acting, so as to avoid unilateral radial forces and to minimize the dimensions.

The only difference between the engine and the pump is that the former has no regulating cylinder.

This type of gearing is made by the Schwartzkopff Engineering Works in standard construction up to a capacity of 200 H.P., but some have also been made already for larger powers of 350 and 400 H.P. and even for 1000 H.P.

Trials carried out at the Technical High School of Dresden have shown that the efficiency of this gearing for a capacity of 50 H.P. is 70—75% at low speed and 80—83% at high speed (see figs. 26 and 27). The optimum temperature for the oil was found to be 40—60° C. (100—140° F.).

Lauf-Thoma Gearing.

The Lauf-Thoma transmission is made by the "Magdeburger Werkzeugmaschinenfabrik A.G." of Magdeburg. It consists of a piston pump and a piston engine (figs. 28a, b, c, d). The pump has radially arranged cylinders, draws oil from a tank and delivers it to a similarly designed piston engine. The volume of oil and consequently the number of revolutions of the piston engine are regulated by adjusting the length of stroke of the pistons, the oil pressure thereby adjusting itself to the power to be transmitted.

The pulley driving the pump is coupled to the rotating cylinder casing, in radial bores of which the pistons are moving, guided by large crossheads. The pistons are supported by rollers on a rotating cylindrical guide, which by means of a handwheel can be adjusted eccentrically in relation to the cylinder casing, thus regulating the stroke of the pistons. The journal around which the cylinder casing rotates has

Fig. 25. Schwartzkopff-Huwiler Gearing. Schnitt = section.

several bores through which the oil flows. During the suction stroke, that is to say while the piston is moving outwards, one half of these bores are connected with the various cylinders by corresponding distributing openings. In the dead centre the pistons are cut off from the oil supply pipes by distributing dams. On the next inward stroke of the cylinder the oil is delivered to the piston engine via other distributing openings and bores. A sufficient number of pistons ensures a continuous flow of oil.

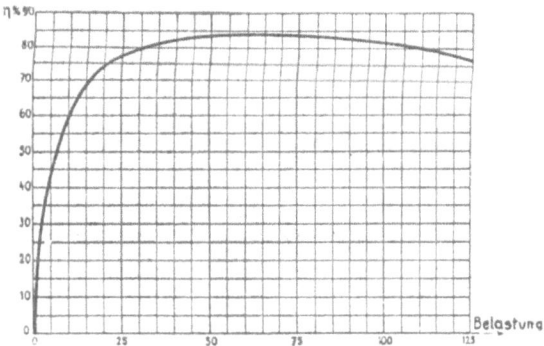

Fig. 26. Efficiency of Schwartzkopff-Huwiler gearing.
Belastung = load.

Other systems.

Among others there are also the *Janney* system [1]) and the *Hele Shaw* system [2]), both acting with piston pumps.

ELECTRIC TRANSMISSION.

Electric transmission has already reached a far advanced stage of development and is very much applied. In this system a dynamo is coupled to the combustion engine and the power generated is supplied to one or more electric motors coupled to the driving wheels of the traction vehicle in the same way as with electric locomotives or motor vehicles.

With a view to reducing the dimensions and the weight of the combustion engine and the coupled dynamo it is advisable to choose a high engine speed. For high powers, however, — which are required only for locomotives for which exclusively Diesel engines can be considered — the engine speed is generally limited to 500—600 revs./min, owing to the system of fuel injection and the combustion process. It is also of importance to keep down the weight of the electric motors, which can very easily be done by installing several small high-speed motors, the power of which can be transmitted to the driving axles of the vehicle by gearwheels.

This leads to the application of axle-motors, each driving axle being driven by one or two motors, which can be mounted in various ways, viz.:

a) rigidly fixed to the frame, allowance being made for some play in the transmission between the motor and the axle;

b) in the manner adopted for tram-cars, where the motor is suspended on hinges from the frame with one end bearing on the driving axle.

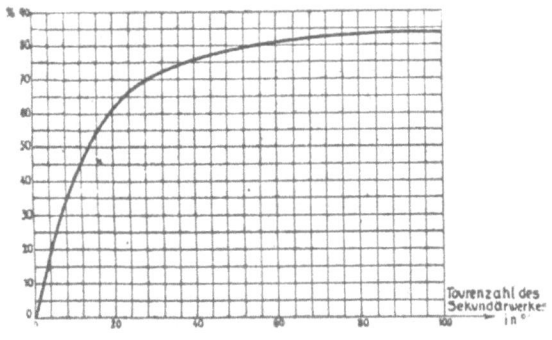

Fig. 27. Efficiency of Schwartzkopff-Huwiler Gearing.
Tourenzahl des Sekundärwerkes = speed of secondary gearing.

[1]) See Engineering 1925, page 13.
[2]) „ The Engineer 1913, page 154.

The most suitable is direct current, for this allows of simplicity in the connections and electric apparatus, and moreover, is most reliable.

In this case there is no need of compressed air for starting the combustion engine, as this can be effected, at least for low powers, by using a battery — which has to be installed in any case for lighting purposes — for running generator coupled to the engine shaft, as electric motor for starting the combustion engine until ignition is brought into action, after which the generator can be switched over, the engine running light until the normal speed is reached.

For higher power units too large a battery would be needed for starting the combustion engine, and in such cases, therefore, the exciter is mostly driven by a special small combustion engine, by means of which the exciter-dynamo can be run as a generator and the power generated can be supplied to the main generator acting as motor until ignition takes place in the combustion engine proper.

The great advantage of the electric transmission lies in the possibility it offers of regulating the speed of the vehicle in any way desired both forward and back-

Lauf-Thoma Gearing.

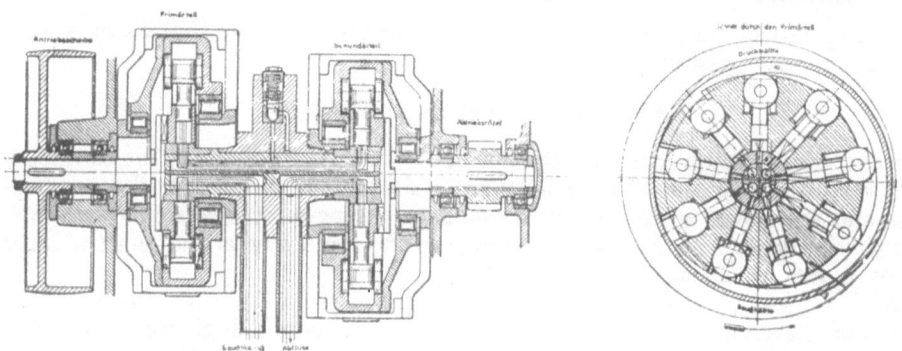

Fig. 28a. Longitudinal section.   Fig. 28b. Cross section.
Abfluss = Outlet. — Antriebsritzel = Driving pinion. — Antriebsscheibe = Driving pulley. — Druckhälfte = Discharge part. — Primärteil = Pump part. — Saughälfte = Suction part. — Saugleitung = Suction piping. — Schnitt durch den Primärteil = Section through pump part. — Sekundärteil = Engine part.

ward, which is arranged in a very simple way. The operation is also simpler than that with any other method. On the other hand there is the drawback of heavy weight and higher initial cost, as will appear from the examples described hereafter.

A special type of electric transmission for Diesel electric locomotives is that of *Krupps*. There are two Diesel engines, one acting direct on the driving axles and the other through the intermediary of a generator and electric motor. The latter engine sets the locomotive in motion until the direct-coupled engine starts working, after which the engine with electric transmission is set for constant power, the tractive force and speed being regulated further with the direct-coupled engine by regulating the fuel supply.

There are no great differences between the various types of electric transmission, axle driving motors being used almost exclusively.

The switch systems, however, differ rather considerably. In the main there are two systems of switching, viz.:
a) where the speed of the combustion engine remains constant, and
b) where this is not the case.

With the first system the speed of the engine is kept constant by means of a centrifugal governor, while the tractive force and the speed of the vehicle are regulated by varying the electric field of the generator. This can be done in a manner which became of practical importance when it was applied for the first Diesel electric locomotive built for Russia.

The exciter derives its power from a continuous and adjustable tension set up, in the case of the Russian Diesel locomotive, by a special auxiliary exciter. This exciter aggregate was driven by a special combustion engine with constant speed. The disadvantage of this system is that the Diesel engine does not run economically, its efficiency depending for a great deal upon the mean indicator pressure, which must necessarily fluctuate considerably if the speed of the engine is kept constant. Furthermore this system requires a second exciter, a main controller, regulating resistances and measuring instruments, all of which call for great skill and care on the part of the driver. Moreover the system does not ensure smooth working, much depending on the skill of the driver whether the starting is shock free or not.

With the second system of switching in addition to the variation of the electric field of the generator also the speed of the combustion engine and thus also that of the generator is varied. One of the best known applications of this system, which is much used in America, is that of the "Forges et Ateliers de Construction Electriques de Jeumont" and of the "General Electric Company" (*Lemp's* switching). In the French design the generator is excited by a self-exciting dynamo driven by the driving axle of the vehicle. The advantage of this construction is that it ensures an absolutely smooth variation of tractive force when starting. Only one exciter is required and switching is done by means of a simple crank, whilst moreover the efficiency of the Diesel engine under partial load is much more satisfactory. A disadvantage,

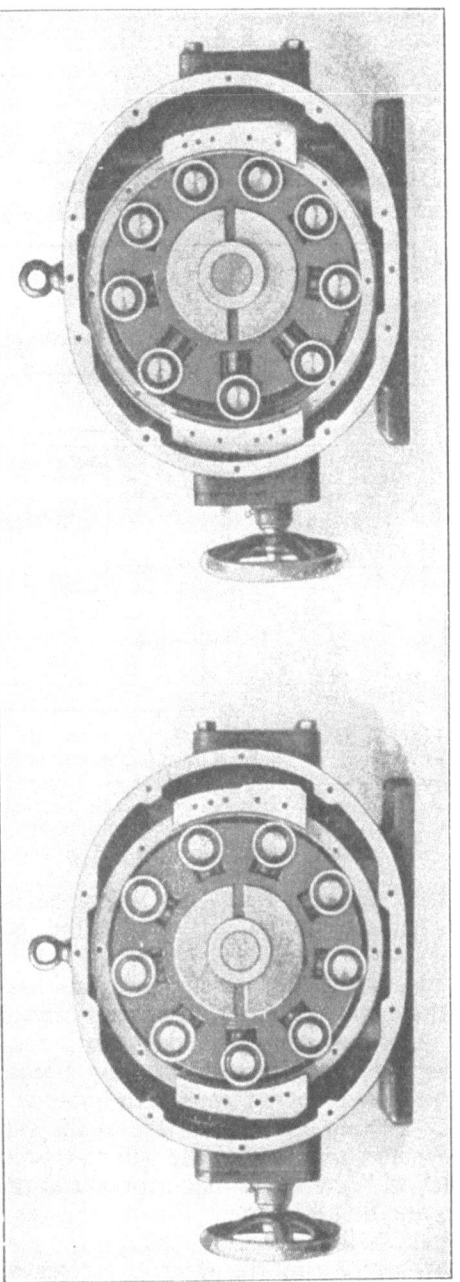

Fig. 28d. Drum in eccentrical position.

Fig. 28c. Drum in concentrical position.

however, lies in the driving of the exciter from the axle of the vehicle, because if anything should happen to that axle the engine would stop at once.

With the *Lemp* switching (fig. 29) the generator and the exciter are coupled to the combustion engine. In addition to the exciting winding the generator has also a reversing pole and a differential compound winding through which the current of the motor flows. The electric field of the generator is excited by the exciter-dynamo, which is itself excited by an accumulator battery. Therefore, whilst embodying the same advantages as those of the French system this manner of switching eliminates the disadvantage of the separate drive of the exciter-dynamo from the axles of the vehicle. Furthermore this system works quite smoothly, without any measuring apparatus or main controller, only by regulating the Diesel engine.

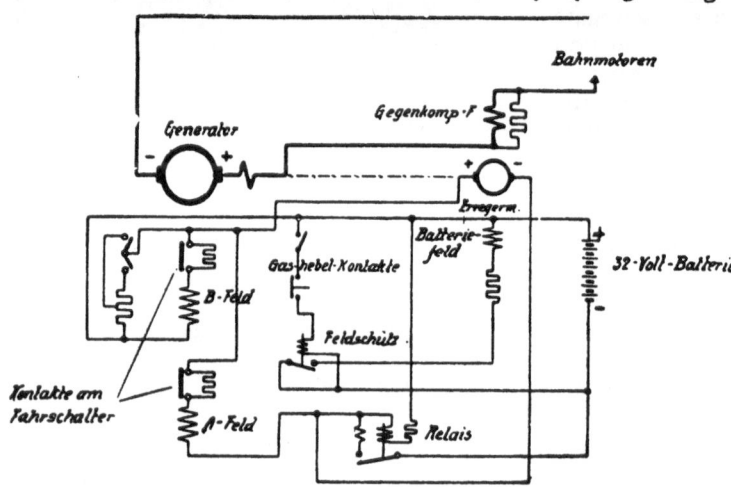

Fig. 29. Diagram of the Lemp automatic control gear.
A-Feld = A-Field. — Bahnmotoren = Tractions motors. — Batteriefeld = Battery field. — B-Feld = B-Field. — Erregerm = Exciter. — Feldschütz = Field safety-coil. — Gas-hebel-Kontakte = Gashandle contacts. — Gegencomp. = Differential compound winding. — Relais = Relay. — Kontakte am Fahrschalter = Contacts of main controller.

The *Gebus* system of Messrs. Gebus-Lokomotiven, Vienna, is a kind of electric transmission with which the power of a combustion engine is transmitted to the driving axles of a vehicle in such a way as to answer, within wide limits, to the required tractive force without needing special attention on the part of the driver; that is to say an increase of tractive force will automatically give a decrease in speed corresponding to a constant engine power. When, for instance, the resistance is increased to twice its original amount the speed will be halved, whilst the power of the combustion engine remains the same.

With the *Gebus* system only one handle is needed for regulating the speed; this is done by varying the fuel supply and thus the speed of the engine. The voltage of the generator changes directly with the variation of speed. At its lowest speed the generator is not excited and only a small residual current will flow through the reversing switch to the connected motors. By increasing the charge and thus the speed of the unit and the voltage of the generator a current will be supplied to the electric motors and the vehicle will start smoothly. The next increase in speed of the generator will give a still higher pressure; the latter causes a stronger current and the vehicle will be accelerated. Finally the required tractive force will be balanced by the corresponding speed.

During the run the electric motors will take up a current corresponding to the resistance. To allow for a constant load of the engine the voltage of the generator has to vary inversely to the current. The product current × voltage will be practically constant within the required limits and the driving engine will be practically constantly loaded when running at a constant speed. When running at a different speed its power will vary proportionately, and the torque will be practically the same. This ensures the best possible working and a minimum fuel consumption.

With a *Gebus* vehicle the combustion engine has to be started first, which is done with the generator, which for this purpose has to be provided with a starting winding. Then the reversing switch has to be placed in the proper position and the speed of the set will be increased until the generator is not only excited but also supplies a current for starting the vehicle. This starting and also the acceleration is done without any shocks and no speed variation is required until full speed is attained. The speed regulation is effected by varying the fuel supply, likewise without any shocks.

A suitable engine design will eliminate all fear of critical speeds; the lower critical speeds will have to be below that at which excitation takes place, and the higher ones must be higher than the normal speed. The harmful effects can be minimized especially by using vibration dampers.

To stop the vehicle the inlet of the engine has to be throttled to such an extent as to make the engine run at its no-load speed, and the excitation will cease. Then the brake can be applied. For long runs down gradients the set can be stopped.

There is no fundamental difference between explosion and constant pressure operation with the *Gebus* system, with which the constant pressure engine can be regulated as usual for these engines.

The main advantages of this system are its simplicity, the completely automatic speed variation of the vehicle and the easy governing, as the speed of the vehicle can be changed simply by shifting the fuel-supply handle from zero into its farthest position.

KITSON-STILL COMBINED DRIVE.

The Kitson-Still locomotive designed by Col. *Kitson-Clark* of Kitson & Co., Ltd., Leeds, is a combination of a steam and a Diesel locomotive. The heat of the combustion gas of the Diesel engine, which is otherwise lost, is utilised for the generation of steam. The water cooling jackets of the engine cylinders are connected up in a closed system with a boiler fired with the combustion gas, whilst the steam is used in special steam cylinders or in the Diesel engine cylinders underneath the pistons.

The fact of the temperature of the water in the cylinder cooling jackets being kept considerably above its boiling point avoids the necessity of raising the compression to the high pressure required in ordinary Diesel engines.

The main advantage of this system, however, lies in the use of steam for starting up, whilst once the locomotive is set in motion it is further propelled by the Diesel engines, and as starting is done with steam a direct transmission is possible from the engine cylinders to the driving axles. However, in order to attain the highest possible engine speed the arrangement is such that the engine drives a blind axle, which works on the driving wheels via a gear wheel transmission and coupling rods.

SCHNEIDER TRANSMISSION.

The *Schneider* transmission [1]) is a combination of a mechanical and an hydraulic

---

[1]) S.B.Z. 1925 I, pp. 123/154.
    Z.V.D.I. 1925, pp. 499/595.

system. The power is transmitted partly mechanically (by gear wheels) and partly hydraulically.

Starting is done hydraulically, followed by a gradual change over to mechanical transmission, so that at normal speed the transmission is entirely mechanical.

The advantage is that starting can be done gradually and smoothly, whilst in normal service the transmission is mechanical and the efficiency high.

SUMMARY ON TRANSMISSION.

Before deciding as to which system of transmission is to be preferred under certain circumstances various factors typical of the systems have to be considered, the principal of which are:

a) reliability in working,
b) smoothness in variation in connection with the speed and tractive force required,
c) weight of the aggregate and price,
d) efficiency.

With *mechanical transmission* the variation of speed is accompanied more or less by shocks. Particularly when starting, thus when the motion of the running engine has to be transmitted to the stationary driving axles, provision has to be made for the smoothest possible working by means of a slip coupling. It is also advisable to have a slipping or springing element for changing over from one gear to another, so as to avoid shocks as far as possible.

When changing speeds the transmission between combustion engine and driving mechanism must be uncoupled from the first set before being coupled up again to the second set of gear wheels. During this changing over no power is transmitted to the driving wheels and this may be troublesome when starting up or when climbing a gradient.

The greater the power of the engine the longer it will take to change over, so that under the heaviest working the conditions will be most unfavourable.

From the foregoing it will be realized that in many cases, owing to the unsatisfactory utilisation of power and the temporary cessation of the tractive force when changing speed, and further the difficulty of obtaining a suitable friction coupling, power transmission from combustion engine to driving axles by a gear wheel transmission with various transmission ratios is in many cases unsatisfactory, and particularly so for shunting, for goods and passenger train locomotives which frequently have to start under heavy load, and for mountain locomotives with considerable variations of tractive force.

On the other hand mechanical transmission is very suitable for express locomotives in even country when the train load can be arranged so that the maximum tractive power of the Diesel engine is not called upon until the highest speed is reached and can be utilised on longer runs. In any other case mechanical transmission seems to be unsuitable.

Mechanical transmission is, however, the simplest and is particularly suited for small units of lower power. So far the Russian 1200 H.P. Diesel locomotive is the only large unit fitted with mechanical transmission.

The efficiency of mechanical transmission varies between 85 and 95%.

With *hydraulic transmission* a liquid is circulated by a pump driven by the combustion engine and this liquid is used to drive the driving axles by means of one or more hydraulic turbines. In general the objection against this system is that the flow passages are as a rule relatively small, resulting in high velocities and heating of the liquid, with consequent losses; the same results from the continual change of the direction of flow.

Consequently the efficiency of the hydraulic system of transmission is not very high, as a rule no more than 80%, but mostly between 65 and 75%.

A smooth change of speed is, however, obtainable. The maximum power that can be transformed by hydraulic transmission is 400 to 500 H.P.

Pneumatic transmission has the disadvantage that the compressed air supplied by a compressor driven by the combustion engine increases considerably in temperature, thus necessitating cooling from the outside, whilst compression has to take place in stages so as to avoid burning of the lubricating oil. As quickly as the temperature rises during compression, so it falls during expansion in the cylinders driving the driving wheels of the vehicle, thus making reheating during the expansion desirable.

This system of transmission has not yet been used very much, so that little is as yet known of the results obtained with the various forms of construction.

Electric transmission is most suitable for medium and high power traction vehicles, though there is the disadvantage of great weight and high cost; the electric transmission for large locomotives weighs 25 to 35 kg (55—77 lbs.) per H.P. whilst a Diesel electric locomotive costs more than twice as much as a steam locomotive of the same capacity.

The efficiency of electric transmission is about 85%.

III. TYPES OF LOCOMOTIVES ALREADY BUILT.

The locomotives driven by combustion engines may be divided into two classes, viz.:
a) train locomotives,
b) shunting locomotives, also called locomotors,
Train locomotives, which are usually of rather high power, are built exclusively with Diesel engines; petrol engines have never been used for this purpose in view of the high price of the fuel. For shunting engines the cost of fuel is not such an important factor and consequently these are also built with petrol engines.

For indicating the type of a locomotive in respect to the wheel arrangement there is a system in use in a large part of Europe by which the pilot or trailing truck wheel sets are indicated by figures and the driving wheel sets by capital letters, a single driving wheel set being indicated by the letter A, two sets by B, three by C and so on. Driving wheel sets which are not joined by driving rods but driven separately are indicated by the index $_0$ behind the letter. For example a locomotive with a pilot truck wheel in front, five driving wheels not joined up by coupling rods and then a trailing truck wheel behind, is indicated as a

$$1 \ E_0 \ 1 \text{ locomotive.}$$

A locomotive with a pilot bogie of two wheel sets in front and behind, with three sets of driving wheels in between coupled up by coupling rods, is called a

$$2 \ C \ 2 \text{ locomotive.}$$

(In England and America these examples would be termed 2—10—2 and 4—6—4 locomotives respectively.)

FIRST SULZER LOCOMOTIVE.

The first train locomotive with Diesel engine was completed in 1912 by Sulzer Bros. in cooperation with *Diesel* and *Klose* for the Prussian-Hessian State Railway. It has already been fully described in various journals [1] and from figs. 30 and 31 attached it will be seen that it is a 2 B 2 (4—4—4) type of express locomotive. The drive consists of a loose axle with two cranks at an angle of 180°, on which the piston rods of the four engine cylinders work. The cylinders are placed in pairs at an angle of 45° from the vertical, thus working on the cranks at an angle of 90° to each other. The engine is a two-cycle one, so that in one revolution of the crank shaft each cylinder piston makes one working stroke, and as these strokes follow each other at 90° the torque is fairly even.

---

[1] ZVDI 1913, p. 1325.
SBZ 1913, p. 297.
Engg 1913, p. 317.
GA 1914, p. 127.
PMK 1914, No. 1.

The direction of rotation and the charge of the cylinders are changed by means of an eccentric fitted on the crank shaft and controlled from the driver's cabin at either end of the locomotive coach.

The air required for combustion is supplied by compressors driven by the crank shaft, but as these compressors are put out of action as soon as the engine stops an auxiliary compressor has been provided, which is driven by a separate Diesel engine and compresses air in air vessels. In this way the locomotive can be started with compressed air until its speed is high enough to set the main Diesel engine running as driving power; experience shows that this takes place when a speed of 8—10 km (5—6 miles) per hour is reached.

Fig. 30. First Sulzer Diesel locomotive with direct drive.

The cylinders of the main and auxiliary engines are water cooled, and moreover the pistons are cooled by the circulation of oil.

The principal dimensions of this locomotive are:

Power . . . . 1000—1200 H.P.
Tare . . . . 95 tons
Maximum speed . . . . 100 km/h (62 m.p.h.)
Diameter of driving wheel . . . . 1.75 metres (5′ 9″).
Length over buffers . . . . 16.60 „ (54′ 6″).

The results obtained with this first Diesel locomotive did not answer expectations. The main difficulties experienced were in connection with the starting of the train, the main engine having to work with compressed air until ignition took place in the cylinders. This required several revolutions to be made and owing to the repeated expansion of the compressed air the cylinders were cooled off too

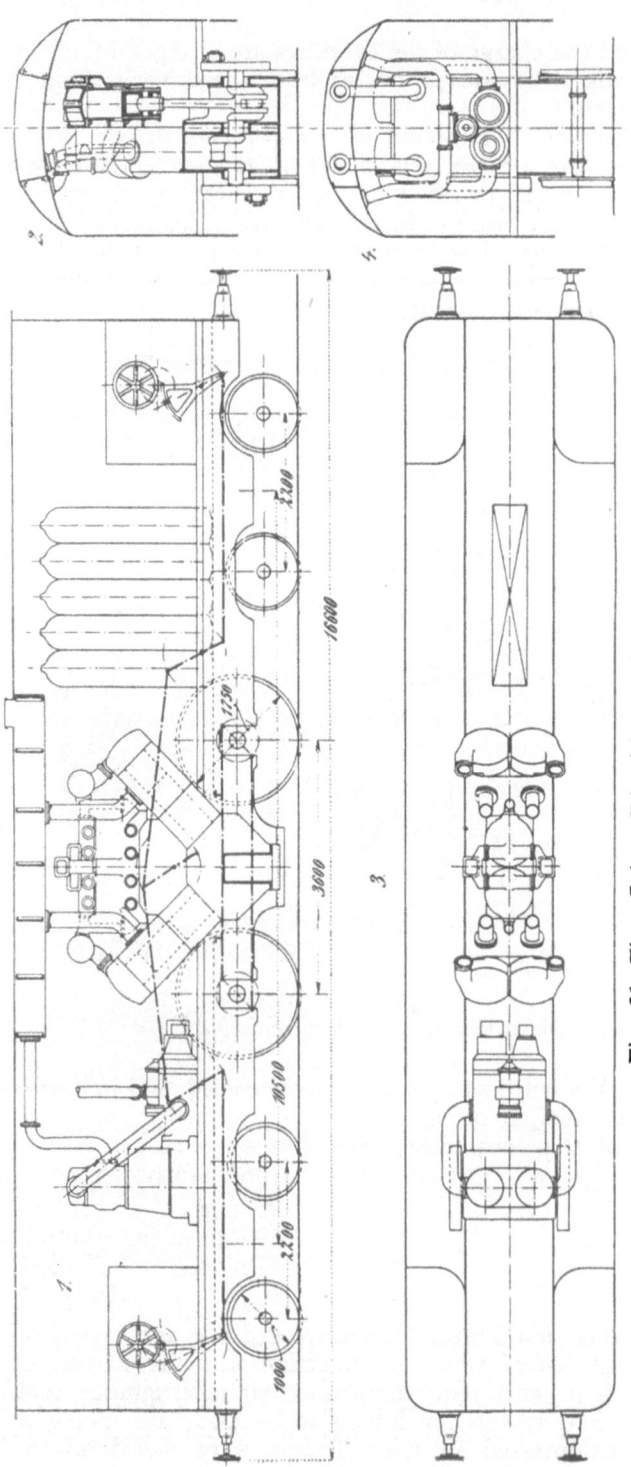

Fig. 31. First Sulzer·Diesel locomotive with direct drive.

much, resulting in imperfect ignition and sometimes violent explosions, which led to breakdowns in the machinery.

These facts lead to the conclusion that direct drive with compressed air starting power is not the most suitable design, and as a matter of fact practically no more locomotives have since been built on these lines.

RUSSIAN DIESEL-ELECTRIC LOCOMOTIVE. 1 E₀ 1 (2—10—2).

This locomotive was designed by *Prof. Lomonossoff*, railway technical adviser to the Russian Government, in collaboration with the "Hohenzollern A.G. für Lokomotivbau" of Düsseldorf; in view of complications arising from the French occupation of the Ruhr district after the Great War the locomotive could not be built at Düsseldorf and the order had to be given to the "Maschinenfabrik Esslingen". A Diesel-electric locomotive was designed in view of the difficulties experienced in some parts of Russia in getting adequate supplies of water suitable for boilers, whilst there was also the consideration that the use of oil engines would provide a good market for the oil produced in the country; a great many steam locomotives in Russia are now oil-fired, the liquid fuel being obtainable anywhere.

In deciding upon the type of locomotive account had

to be taken of some very sharp and long gradients occurring on the lines on which the locomotive was to be used. It was necessary for the locomotive to have a max. tractive force of about 15000 kg (33,000 lbs.) and about 900 H.P. at the periphery of the driving wheel at a speed of 15 km/h (9 m.p.h.); for goods trains a max. speed of 30 km /h (19 m.p.h.) had to be attainable.

Reckoning on a loss of 20—30% in the transmission, a brake horse-power of 1200—1250 was considered necessary. This led to the choice of a six-cylinder four-stroke Diesel engine with a speed of 450 revolutions per minute, similar to those built for submarines by the "Maschinenfabrik Augsburg-Nürnberg" at Augsburg, to which works the order was given for the Diesel engine. The electric equipment was

Fig. 32. Russian Diesel-electric locomotive.

made by Brown Boveri of Baden. The coupling was carried out on the *Ward-Leonard* principle, but with two exciters, which allows of an exceptionally fine adjustment.

Figs. 32 and 33 are illustrations of this locomotive as it was originally built.

For the engine cooling water and lubricating oil two special coolers were installed, with brass pipes, with a total cooling surface of 430 m². Four double ventilators coupled to the engine shaft supplied the necessary air at the rate of 25 m³/sec at an engine shaft speed of 450 revolutions per minute. In winter these coolers are sufficient, but in summer a cooling tender has to be coupled to the locomotive.

The Diesel engine and the direct current generator are connected up together by a flexible coupling and mounted in the locomotive lengthwise. The maximum voltage of the current supplied by the generator is 1000 volts.

The five motors are connected parallel and mounted as axle motors with tram suspension, each motor delivering 200 H.P. for 100 minutes. The gearing ratio is 1: 6.14; the gear wheels are mounted on both sides and made with bevel teeth.

The main dimensions of the locomotive are as follows:

Gauge	1524 mm (5')
Max. tractive force at driving wheels	15000 kg (33000 lbs.)
Speed at max. tractive force	15 km/h (9 m.p.h.)
Max. power at engine shaft	1200 H.P.
Max. speed	55 km/h (34 m.p.h.)
Max. hauling power on gradient 5 : 1000	1900 tons
,, ,, ,, ,, ,, 10 : 1000	1000 ,,
Adhesion weight in working order	95 ,,
Total weight ,, ,, ,,	120 ,,
Fuel bunker capacity	4 ,,
Number of cylinders	6
Diameter of cylinders	450 mm (17³/₄'')
Piston stroke	420 ,, (16½'')
Revolutions per minute	450

The locomotive was completed in June 1924, when it was tried out on the testing bed and compared with a Russian steam locomotive of the same power with five coupled axles and superheater; this steam locomotive was fired with the same oil used for the Diesel engines. The results of these comparative tests are tabulated below:

Speed		Tractive force in tons		Total efficiency (from fuel to draw-hook)	
km/h	m.p.h.	Steam loc.	Diesel-electr. locomotive	Steam locomotive	Diesel-electr. locomotive
0	0	15.3 [1]	19.1 [1]		
10	6	15.3 [1]	19.1 [1]	7.6	24
20	12½	14.0	13.3	8.1	27.5
30	18	9.8	9.2	8.5	28.5
40	25	7.8	6.8	8.6	26.0
50	31	6.7	4.0	8.5	25.2

[1] Max. value determined by adhesion.

After the trials were finished some alterations were made: the auxiliary engine for the exciter was removed and the exciter coupled direct to the generator by means of gear wheels, which raised the total efficiency from 26 to 29%; further, in order to save weight, a cooler was removed and the other altered, the cooling water passing through the pipes instead of air and the air circulating around the tubes, this increasing the heat conductivity from 54 to 110 kcal per sq. metre per hour and per degree Celsius temperature variation (from 13.6 to 27.7 B.T.U./h/° F.). In place of the cooler removed, a cooling tender was built with 3 equal cooling elements, each with a double ventilator like that on the locomotive, driven by a three-cylinder Diesel engine of 100 H.P. (figure 34). The advantage of providing such a cooling tender is that it need only be carried in summer, because for winter conditions the cooling element on the locomotive itself is sufficient. No alterations were made to the main Diesel engine as the trials had shown that a balanced Diesel engine may run perfectly and without any unpleasant vibrations.

After these changes had been made fresh comparative tests were carried out on the testing bed with the same oil-fired steam locomotive as used before. The tests had to approximate as near as possible to the actual conditions in railway work. It was decided to make three trial runs A, B and C with each locomotive with a weight of train of 1800 tons (including weight of locomotive). The gradients and other requirements were fixed as follows:

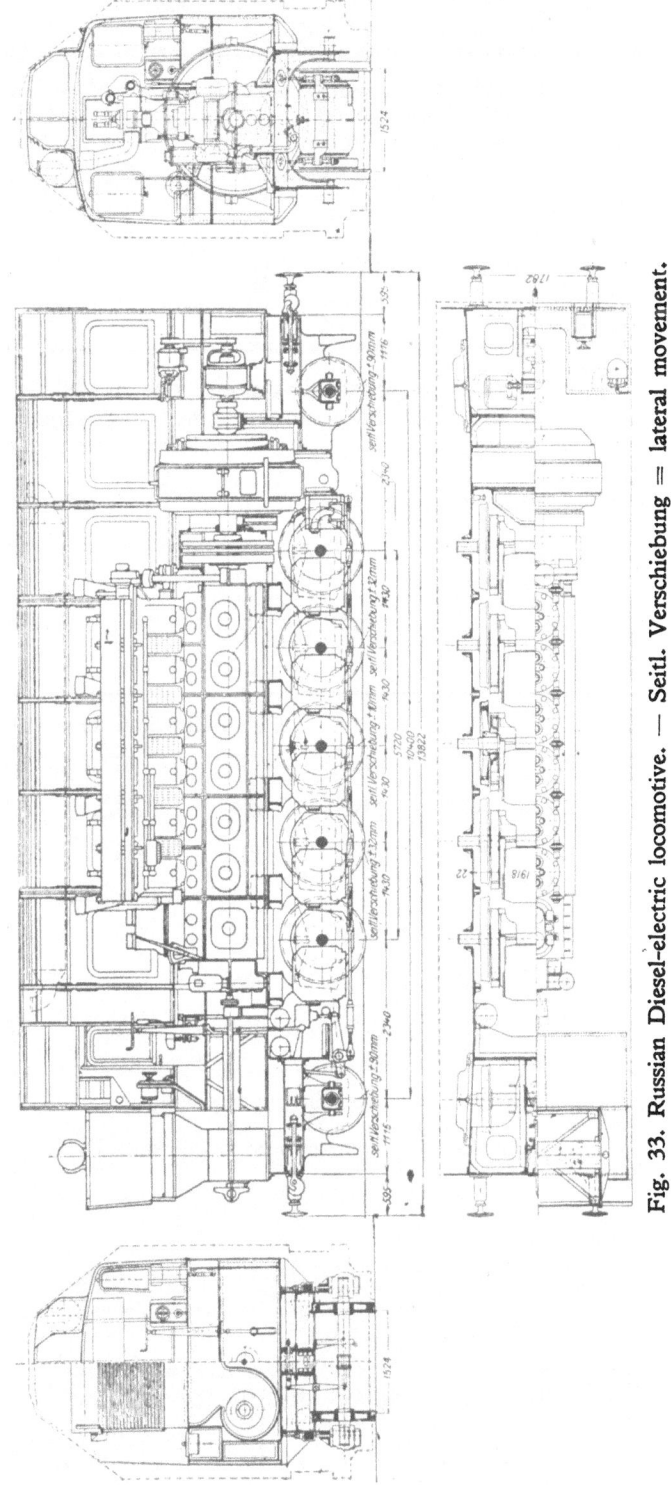

Fig. 33. Russian Diesel-electric locomotive. — Seitl. Verschiebung = lateral movement.

Run	Length of Run in km	Gradient up or down	Speed km/h	Time run in min	Tractive force at driving wheels in kg	H.P.
A	23	1 : 435 up	25	55	9000	830
B	30	1 : 2000 down	30	60	4500	500
C	22	1 : 158 up	14	94	15200	790

Run C conditions were chosen in view of the gradient to be climbed on the line in Russia on which the locomotive was to be run, whilst at the same time this test would give an idea of the performance of the locomotive under full load on a longer run.

The results of the trials are given in the following table:

Run	Time in min	Avge tract. force in kg	Avge speed km/h	Avge power at driv. wheel periphery in H.P.	Total fuel cons. in kg	Fuel cons. in kg per H.P.h	Total eff. in %	Cylinder charge	Re-marks
Diesel electric locomotive									
A	55	9100	25.5	857	183.4	0.23	27.4		a
B	60	4325	29.6	474	113.5	0.237	26.5		b
C	94	15240	15.8	890	348[1]	0.25[1]	25.3		c
Steam locomotive									
A	57	9070	24.7	835	566	0.73	8.67	35	d
B	59	4300	29.5	468	388	0.835	7.6	22	e
C	41[2]	14480	17.0	910	514	0.83	7.62	50	f

[1]) incl. auxiliary engine on cooling tender; latter in use for 33 min. and temp. of cooling water reduced to 66° C. (atm. temp. 13° C.)

[2]) Run stopped after 41 min on account of weakening of a spring of one of the cylinder valves.

Remarks:

a. Engine ran 20 min to heat up; cooling ventilator not in use; temp. of cooling water at end of trial 47.5° C. (atmospheric temp. 11° C.).
b. Cooling ventilators not used.
c. Load reduced for 10 min between runs B and C.
d. Boiler pressure at end of run 162 lbs./sq. in. (11 kg/cm²).
e. „ „ „ „ „ „ 112 „ „ (7.6 „ „).
f. „ „ „ „ „ „ 162 „ „ (11 „ „).

The fact that both locomotives were run on the same fuel helped considerably towards obtaining reliable comparisons.

It was remarkable that whereas hardly any vibrations were noticed with the Diesel locomotive the vibrations of the steam locomotive on the test bed were so severe that the membrane of the instrument for measuring the tractice force was damaged.

In the beginning of January 1925 trial runs on the track were begun in Latvia with heavy goods trains, a dynamometer car being coupled in between the train and the locomotive. The efficiency recorded varied between 21 and 28%, the same as on the

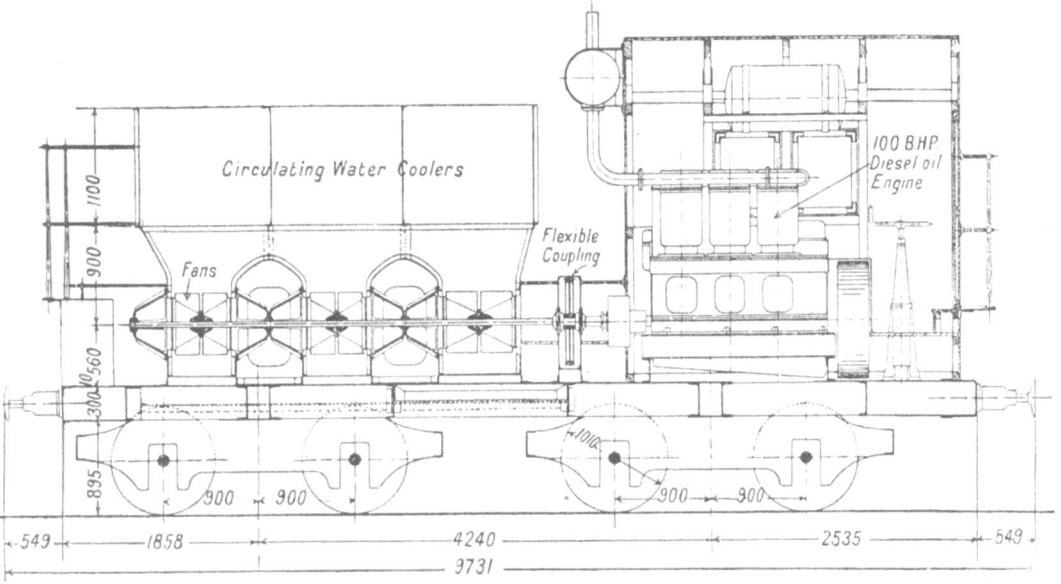

Fig. 34. Cooling tender for the Russian Diesel-electric locomotive.

test bed. Statistics show that the fuel consumption of the Diesel-electric locomotive is $4\frac{1}{2}$ times less than that of steam locomotives of the types E (0—10—0) and 1 D (2—8—0) in service on the same lines.

Another noteworthy point is the exceptional ease of starting with such heavy trains, not only on the level but also on gradients. The last trials were made with a train load of 1950 tons on a gradient of 5 : 1000 at a speed of 15 km ($9\frac{1}{4}$ miles) per hour, when the fuel consumption was found to be 226 grammes (0.5 lb.) per H.P. hour at the periphery of the driving wheel, which means a total efficiency of 28%.

In March 1925 the final trial runs were begun over the Caucasus to the Turkish-Persian frontier. On the return journey, with the exception of one short break, a non-stop run of 2500 km was made. On this journey one of the five motors broke down and as there was not a spare armature available the last 400 km had to be done with 4 motors. The locomotive was eventually taken into a workshop to have the defective armature repaired. At the time this happened the locomotive had already covered 22000 km (13700 miles) and since the armature was repaired it has been continuously in service in Russia.

RUSSIAN 2 E 1 (4—10—2) DIESEL LOCOMOTIVE
WITH MECHANICAL TRANSMISSION.

This locomotive, likewise designed by *Prof. Lomonossoff*, was built by the
"Hohenzollern A.G. für Lokomotivbau" of Düsseldorf. Illustrations are given in
figures 35 and 36, whilst figure 37 shows the locomotive when it was nearing completion.

The Diesel engine is a six-cylinder, four-stroke MAN engine, rotatable in
either direction. This drives, via gear wheel transmission with three speeds, a loose
axle coupled to the driving wheels by coupling rods.

The air compressor is built in at the front end of the engine. An auxiliary
Diesel engine drives the ventilator drawing the necessary air through the water and

Fig. 35. Russian Diesel locomotive with gear wheel transmission.

oil coolers. This engine serves at the same time as reserve for the dynamo and air
compressor, which are driven either by the main engine or by the auxiliary engine
by means of chain transmission. At the front end of the locomotive is the oil cooler
(see figures 35 and 36) and behind that the water cooler. The oil cooler consists
of flat pipes, whilst for the water cooler ribbed tubes are used, air being drawn
through both by the ventilator. There are two more water coolers which can be
put out of service when the atmospheric temperature is low enough.

Between the engine shaft and the gearing is an electro-magnetic friction coupling,
whilst each of the three gear sets is likewise cut in with electro-magnetic couplings;
these friction couplings were made by the "Magnetwerk G.m.b.H." of Eisenach.

Figure 38 is a sectional drawing of the main coupling. At the end of the engine
shaft is a disc A which can be magnetized by a coil B let into a circular groove;
the ends of the coil are fixed to a couple of slip rings at the back of the disc. A heavy
flywheel rim C is fitted to the disc A and to this rim is attached the friction plate
D; cooling channels are provided between the ring C, the disc A and the plate D.

The engine shaft also carries a rotatable disc E bearing a plate F to which a ring of friction material G is attached. When current is passed through the coil B the disc A is magnetized and the ring F is attracted, overcoming the force of the spring bolts with which F is attached to E. Consequently the friction material is pressed against friction plate D, so that the rotation of the engine shaft is transmitted to disc E, which is connected via a spring coupling to the driving shaft of the gear transmission. The coupling can be made to slip more or less according to the regulation of the current in the coil B, so that the engine torque can be transmitted to the driving axles of the locomotive gradually.

The electro-magnetic friction coupling for each of the three speed gears is shown in figure 39. Just as with the main coupling, a disc A is keyed onto the driving shaft and can be magnetized by a current passed through the coil B. The disc C is then attracted and the intermediate friction discs D are pressed together. These friction discs are alternately connected to the disc E and disc A, enabling

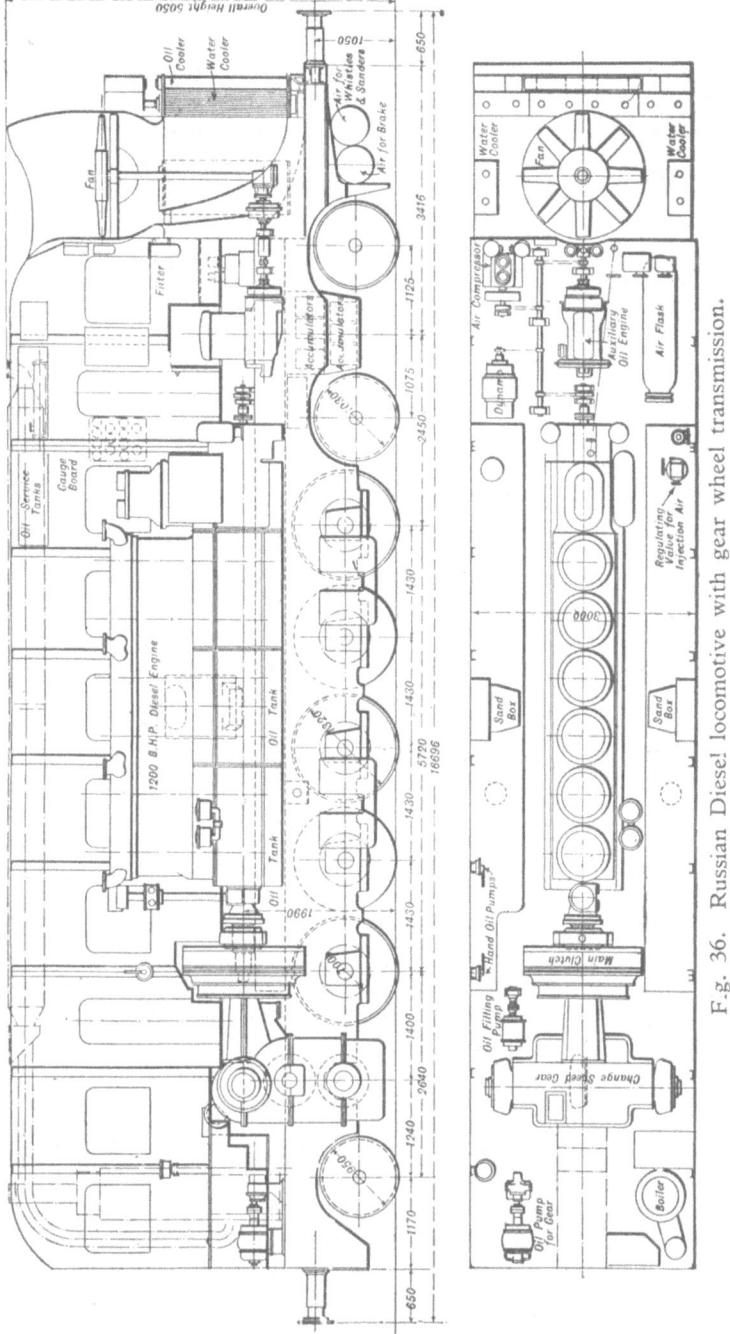

Fig. 36. Russian Diesel locomotive with gear wheel transmission.

Fig. 37. Russian Diesel locomotive with gear wheel transmission during erection.

the rotation of the disc A to be transmitted to a box-shaped disc E rotatable on the driving shaft. These couplings are not designed for lengthy slipping but slip just enough to switch in each of the three gear sets — the teeth of which are continuously in mesh — free of shocks.

Figure 40 shows the change speed gear box for three speeds, viz.: 6.6 : 1—4 : 1 and 2 : 1. This change speed gear was made by Krupp of Essen. As already remarked, the gear wheels are always in mesh and are switched in and out by the electromagnetic couplings described above. The power of the horizontal driving shaft is transmitted to the upper shaft C via bevel gear wheels N, which have a maximum diameter of 500 mm and are capable of transmitting 1200 H.P. at 450 revs./min. The lowest speed is obtained by switching in the right hand coupling I, the other couplings then being switched off. The gear wheel Z_1 is then turned with the shaft C and drives the second shaft B, which in turn drives the loose axle A via gear wheels N. For the middle speed coupling II is put in, turning wheel Z_2 with shaft C and thereby also shaft B, the rotation of which is transmitted to the loose axle A by wheels N. For the top speed coupling III is switched in, which causes wheel Z_3 to turn with shaft B, shaft A again being rotated via wheels N.

Direction of travel is reversed either way by reversing the engine rotation, which is effected in the usual way by bringing the other cams into action for the valve operation.

The main dimensions of this locomotive are:

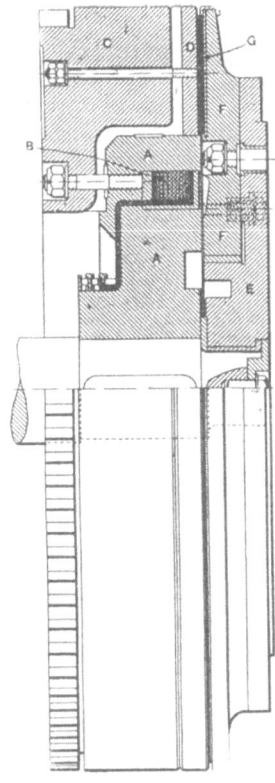

Fig. 38. Main coupling of het Russian Diesel locomotive with gear wheel transmission.

Total weight in working order	131 tons
Adhesion weight	88 tons
Fuel bunkering capacity. .	$3\frac{1}{2}$ tons
Lub. oil storage	1 ton
Water storage	1 ton
Diameter of driving wheels	1320 mm (4' 4")
Max. speed	50 km/h (31 m.p.h.)
Max. tractive force . . .	17.6 tons
Diameter of cylinder . . .	450 mm ($17\frac{3}{4}$")
Piston stroke	420 mm ($16\frac{1}{2}$")
Cylinders	6
Max. revolutions per min of engine shaft	400
Max. power of engine . .	1100 H.P.
Capacity of water pump .	48 m³/h (10600 glns.h)
Capacity of oil pump . .	24 „ (5300 glns.h)
Cooling surface of water cooler	530 m² (5700 sq.ft.)
Cooling surface of oil cooler	200 m² (2150 sq.ft.)
Revs./min of ventilator . .	1200
Air cap. of ventilator . . .	130000 m³/h (4,600,000 cub.ft.)
Max. power of auxiliary engine	60 E.H.P.
Output of dynamo	6 kW
Voltage of battery	110 volts
Power of compressor at 800 atm.	6 E.H.P.

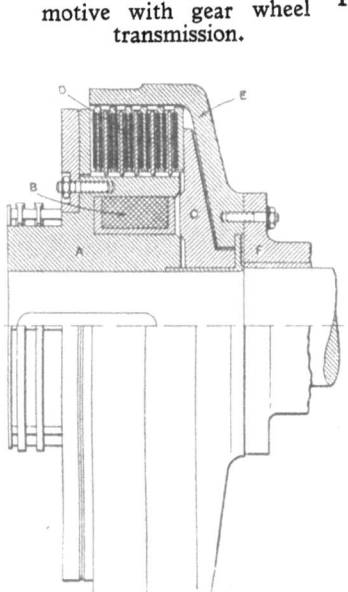

Fig. 39. Friction coupling of the Russian Diesel locomotive with gear wheel transmission.

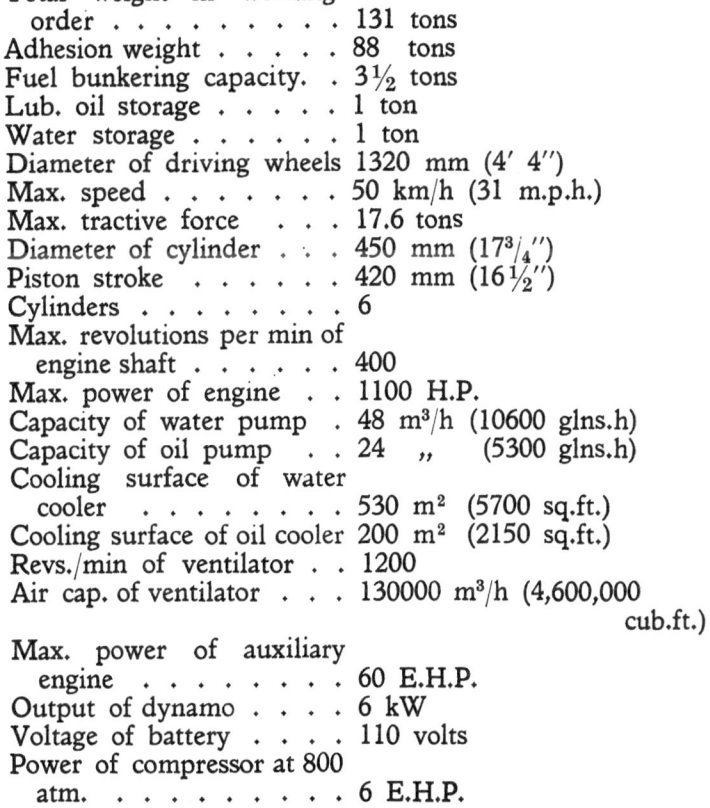

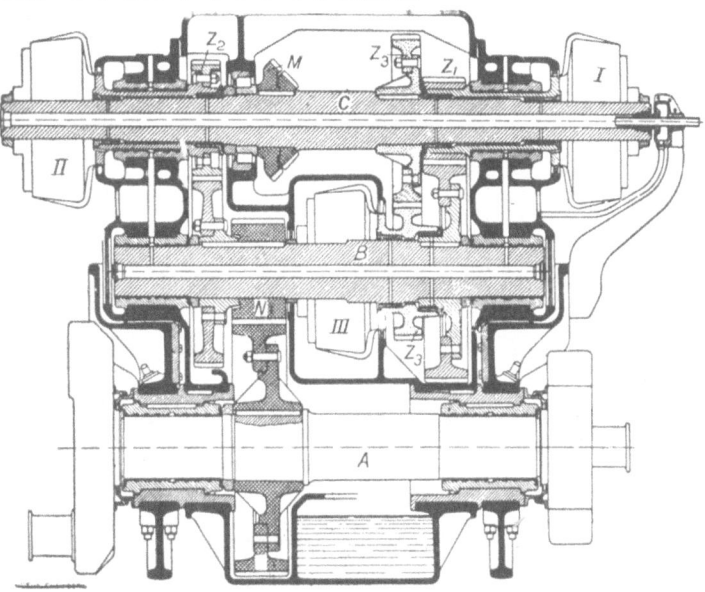

Fig. 40. Gear box of the Russian Diesel locomotive with gear wheel transmission.

RUSSIAN 1 $E_0$1 (2—10—2) DIESEL ELECTRIC LOCOMOTIVE [1]).

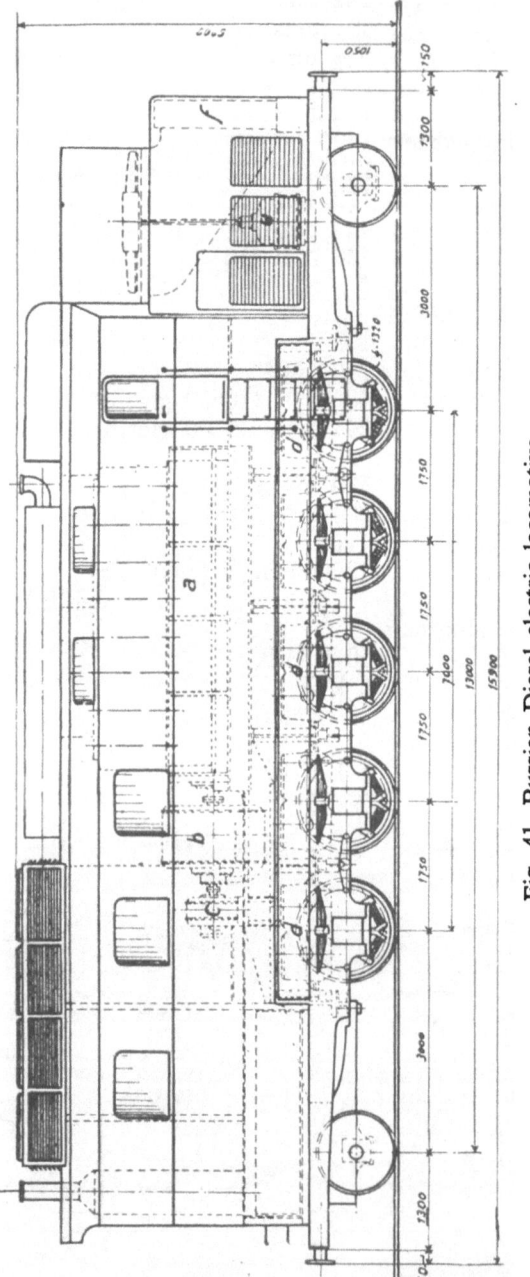

Fig. 41. Russian Diesel electric locomotive.

This locomotive is now in course of construction and will be the heaviest ever yet built. It is destined for service in the arid steppes of South and SE. Russia and is being built by the "Hohenzollern A.G. für Lokomotivbau", Düsseldorf, in cooperation with the Sulzer Works at Winterthur, who are supplying the Diesel engines, and the Sécheron Works at Geneva, the suppliers of the electric equipment.

Figure 41 is a general plan of this locomotive, which is designed for two Diesel generator aggregates totalling 1500 H.P. Diesel engine power. This is the first time that the entirely elastic driving motor suspension with Sécheron single-axle drive has been applied for Diesel locomotives.

The main dimensions are as follows: Tractive force at driving wheel periphery (from start-up to speed of 13½ km/h (8½ m.p.h.)

12000 kg (26000 lbs.)

Engine power
for an hour at driving
wheel periphery 1200· H.P.
Consumption of
total Diesel power up to
45 km/h (28 m.p.h.)
Maximum speed 60 km/h (37 m.p.h.)
Weight in working order
abt. 135 tons
Adhesion weight „ 105 „

The two Diesel engines have a continuous power of 750 H.P. and can be overloaded for an hour up to 825 H.P. The number of revolutions is 650 per min. A generator with exciter is coupled direct to each Diesel engine. The Diesel engines are started by means of the generators, which are provided for that purpose with a special series winding fed by a battery of accumulators. The driving motors are of the Sécheron type and are built in the form of twin motors. The

[1]) Sécheron Mitteilungen, No. 1, 1929, p. 22.

connections are arranged in such a way that in case one Diesel generator group breaks down either the full tractive force can be generated at half speed or the full speed can be attained with about half the tractive force.

DIESEL ELECTRIC LOCOMOTIVE SERIES 2020 OF THE AUSTRIAN RAILWAYS [1]).

This locomotive was built for light local traffic and has been in service since October 1927. Its arrangement is shown in figures 42 and 43 and the main dimensions are:

Gauge	1435 mm (4' 8½'')
Dia. of driving wheel (rim 50 mm thick)	950 mm (3' 1³/₈'')
Length over buffers	10.08 metres (33' 1'')
Total wheel base	4.80 metres (15' 9'')
Driving axle load	12.5 tons
Total weight in working order	37.1 tons
Weight of engine part	8.4 tons
Weight of electr. part incl. battery	11.1 tons
Max. tractive force (short time)	4800 kg (10500 lbs.)
Max. speed	60 km/h (37 m.p.h.)

Fig. 42. Austrian Diesel-electric locomotive.

The Diesel engine is a single acting, six-cylinder, compressorless, four-stroke engine of 200 E. H.P. at 400 revs./min. By increasing the number of revolutions to 460 per min the power can be raised to 225 E.H.P. and for a short time even to 250 E.H.P. The cylinder bore is 225 mm (8⁷/₈'') and length of stroke 340 mm (13³/₈''). Starting can be done with compressed air, for which an air compressor is supplied.

The fuel is drawn by a pump from two elevated tanks of 150 litres capacity each and delivered to the main fuel pump installed for each cylinder. These pumps are regulated so as to keep the engine speed practically constant under all loads. A circulating system of forced lubrication is arranged for the main and crank shafts and steering shaft bearings, by means of an oil pump, whilst the cylinders and compressor are lubricated from a Friedmann lubricating pump. The cooling water is circulated by a pump and cooled back in a double-element cooler fitted with an

[1]) O. Nebesky in "Elektrotechnik und Maschinenbau" 1928, Vol. 52, p. 1193.

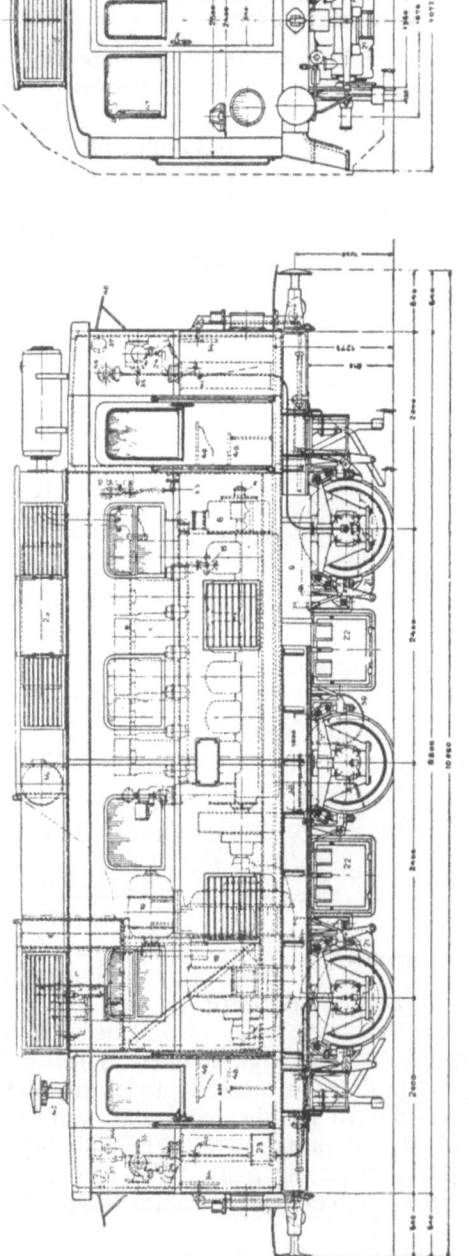

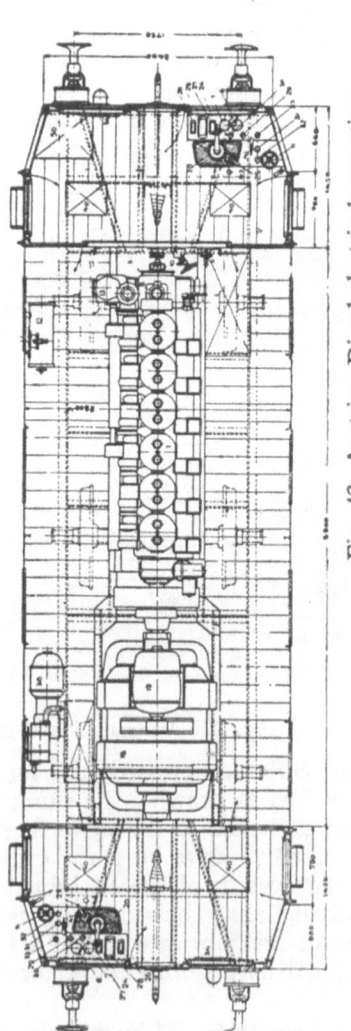

Fig. 43. Austrian Diesel-electric locomotive.

electrically driven ventilator. An elevated reservoir of 130 litres capacity is provided for a reserve of cooling water.

The generator, coupled direct to the Diesel engine, has a continuous output of 160 kW and an hourly output of 180 kW at 600 volts tension, which latter can be increased to 750 volts. As there was no single generator unit available for this voltage and the desired number of revolutions, which is rather small, and gear wheel transmission would involve losses, two normal shunt generators of 300 volts terminal tension each were combined into one double generator (see figure 44).

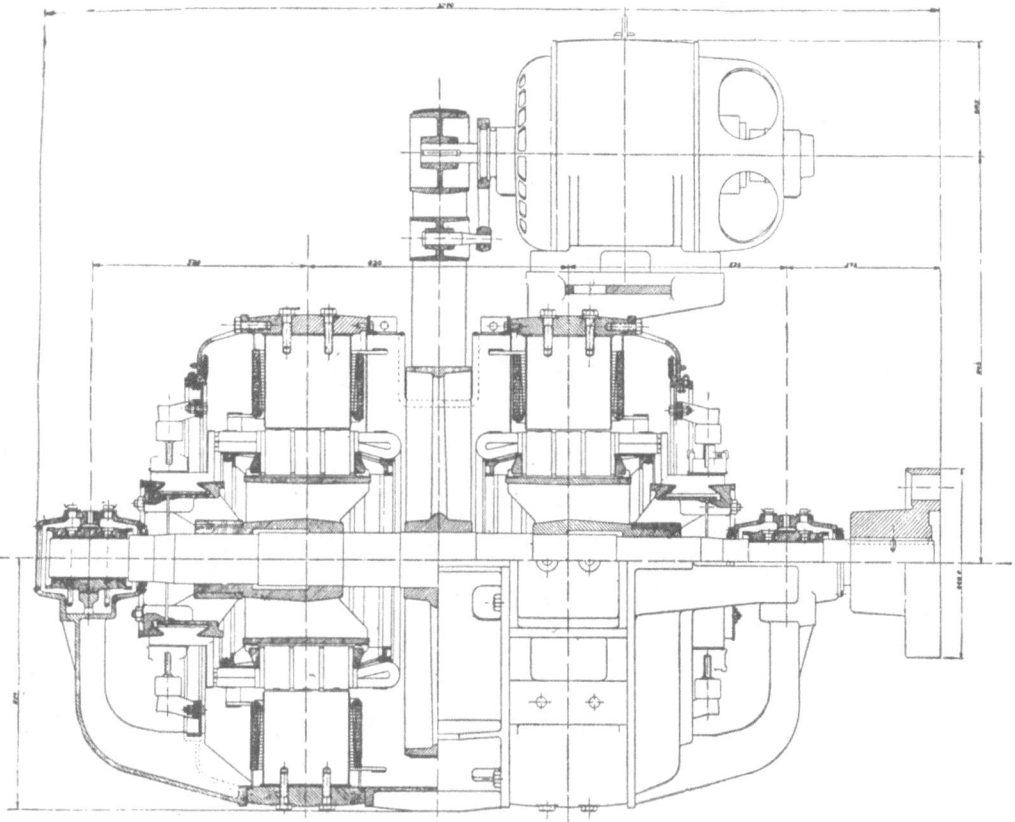

Fig. 44. Cross section through twin generator with exciter. Diesel-electric locomotive of the Austrian Federal Railways.

The Diesel engine is started electrically by running the generator as a motor, the necessary current being supplied by a battery of accumulators. The compressed air installation serves merely as a reserve for starting. By distributing the electric tension between two armatures the advantage of a much smaller battery is obtained, both as regards capacity and in respect to the number of cells, which would have to be greater for a single motor.

A four-pole exciter of 12 kW continuous output at 230 volts tension and 1900 revs./min is driven by the generator via a belt-drive. In addition to the exciting current for the generator this exciter also supplies current for the 6 H.P. motor for the cooling ventilators, for charging the battery of 102 cells and 200 Ah, for the vacuum pump motor, for lighting and for steering.

The two motors act on an end axle of the locomotive and are mounted with tram suspension. The ratio transmission is 1: 5.2. With full field current the hourly output of each motor is 76.5 kW at 600 volts and 700 revs./min and 95 kW at 750 volts and 890 revs./min; the continuous power is 50 kW and 64 kW respectively.

In order to give greater safety the control lever is fitted with a dead-man's push button.

A control cabin is built at both ends of the locomotive (see figure 45) in each of which is a control lever, recording and controlling instruments, a brake handle, a switch for the Diesel engine speed regulator (half speed, normal speed, high speed and stop), a switch for the vacuum brake, handwheel for hand brake, sand-box control, manometer for the lubricating oil and cooling water for the Diesel engine, handwheel for starting controller, push-button for the switch and siren, signal and control lamps and finally light-switches.

Fig. 45. Cabin of Diesel-electric locomotive of the Austrian Federal Railways.

It has been found that the vibrations set up by the Diesel engine are only troublesome when the critical number of revolutions is reached. Under normal speed, and particularly during the run, the vibrations are hardly noticeable.

The locomotive described was put into regular service in October 1927 on the 62 kilometre line Arnoldstein-Kötschah-Mauthen (Gail valley railway). It runs on an average 220 kilometres per day.

Running-out tests showed the resistance of the locomotive at 40 km/h to be between 3.5 and 4 kg (7.7 and 8.8 lbs.) per ton. On one of the trial runs four four-wheeled coaches were taken along, the total train weight being 100 tons. On the return journey the weight was increased to 145 tons by adding two loaded goods trucks, with which load the train was started on a gradient of 13:1000.

The fuel consumption was 5.93 grammes per ton per kilometre (0.021 lbs. per ton mile).

The locomotive was built by the "Grazer Waggon- und Maschinenfabrik

Fig. 46. Narrow gauge Diesel-electric locomotive of the Austrian Federal Railways.

A.G.'', formerly Joh. Weitzer, of Graz, who also built the Diesel engine and made the mechanical parts. The electric equipment was supplied by the "A.E.G. Union Elektrizitäts Gesellschaft'' of Vienna.

The same engineering works at Graz have built another Diesel-electric locomotive for the Austrian Railway, likewise with 200 E.H.P. but for a gauge of 760 mm (2' 6"). This locomotive is illustrated in fig. 46.

SCHWARTZKOPFF DIESEL LOCOMOTIVES.

The "Berliner Maschinenbau A.G. vormals L. Schwartzkopff" have built two Diesel locomotives, which were exhibited at the Railroad Technics Exhibition at Seddin.

Fig. 47. Schwartzkopff Diesel locomotive with pneumatic transmission.

One of the locomotives (fig. 47) is provided with pneumatic transmission, and the main dimensions of this locomotive are:

Engine { power . . . . 220 H.P.
{ revs./min 500
Max. tractive force . . . . 3600 kg (8000 lbs)
Max. speed . . . . 50 km/h (31 m.p.h.)

It can be used for shunting work or for local traffic.

Power is transmitted from the engine to the driving axles on a system patented by the "Waggon- und Maschinenbau A.G.", Görlitz. The exhaust gas is drawn into a compressor, where it is compressed to 8—10 atm. (120—150 lbs.), which pressure can be increased to 15 atm. (220 lbs.) when starting up.

Thus the Diesel engine and compressor take the place of the boiler in an ordinary locomotive. The compressed exhaust gas (temperature 350° C., or 662° F.) performs the same function as superheated steam in a steam locomotive, and consequently the operation is the same. Steam distribution is regulated by a screw handle, whilst the gas throttle is operated by turning a lever. A driver's cabin is placed at both ends of the locomotive. While the locomotive is in service the Diesel engine has to be kept running continuously at the normal speed and when fully loaded consumes 200 grammes (0.44 lb.) fuel of 10,000 kcal (18,000 B.T.U. per lb.)

per H.P. hour. With a view to reducing the power required for driving the compressor the suction gas is cooled before and during compression and only afterwards heated again by the exhaust gas from the Diesel engine.

The cooling water for the Diesel engine and the compressor is cooled by natural and by artificial ventilation. A ventilator driven by the shaft of the engine draws in air and blows it against the cooler mounted on the roof of the locomotive.

Both hand and air brakes are provided.

The Diesel engine is started with compressed air.

After performing its work in the cylinders the exhaust gas of the locomotive is drawn up and compressed again by the compressor.

The other locomotive (fig. 48) has an hydraulic transmission of the Schwartz-

Fig. 48. Schwartzkopff Diesel locomotive with hydraulic transmission.

kopff-Huwiler system. The Diesel engine is likewise of 200 H.P. but with 440 revs./min.

With the exception of the transmission this locomotive is, on the whole, the same as that described above.

1 B (2—4—0) DIESEL LOCOMOTIVE WITH LIQUID TRANSMISSION.

This locomotive, which was built by the "Berliner Maschinenbau A.G., vormals L. Schwartzkopff", Berlin, is shown in figs. 49 and 50.

The engine is a six-cylinder, four-stroke Diesel engine of Görlitz, with compressor for starting and fuel injection. Power is transmitted from the engine to the driving axles by means of the Schwartzkopff-Huwiler system of liquid transmission. The regulating sleeve is operated by compressed air. The Diesel engine itself is not reversible. For changing the travelling direction a rotating valve is provided by means of which the flow of the liquid is reversed.

There are eight coolers, each with a ventilator for cooling the water and the gear oil, the temperature of the latter governing to a certain extent the efficiency of the locomotive. These coolers have yielded favourable results; in the trial runs the highest temperature of the cooling water in the cylinders was 71° C. (160° F.) whilst the average was about 62° C. (144° F.), at an atmospheric temperature of 27—28° C. (80—82° F.).

A special quick running, four-cylinder combustion engine drives the ventilator and cooling water pump. This allows of the main engine being shut down without

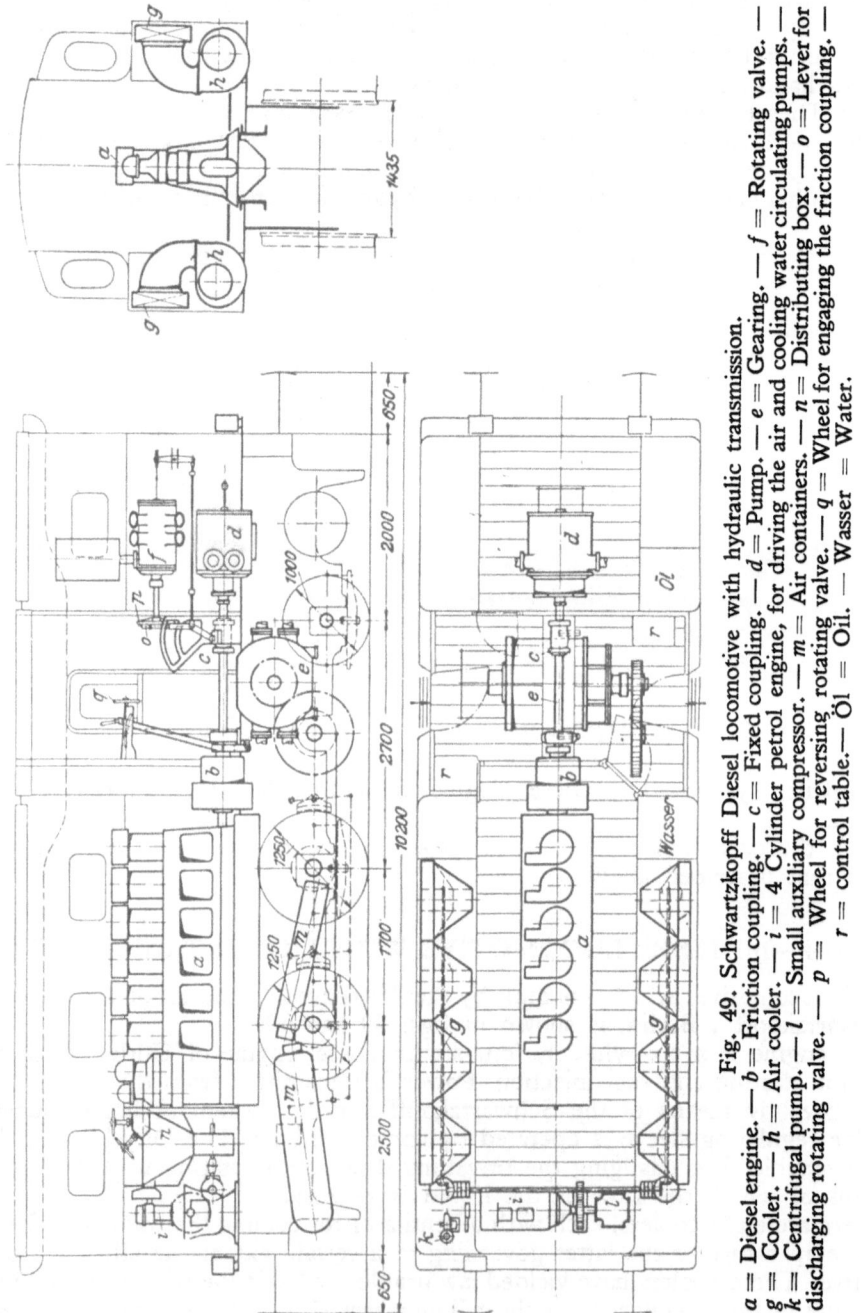

Fig. 49. Schwartzkopff Diesel locomotive with hydraulic transmission.

a = Diesel engine. — b = Friction coupling. — c = Fixed coupling. — d = Pump. — e = Gearing. — f = Rotating valve. — g = Cooler. — h = Air cooler. — i = 4 Cylinder petrol engine, for driving the air and cooling water circulating pumps. — k = Centrifugal pump. — l = Small auxiliary compressor. — m = Air containers. — n = Distributing box. — o = Lever for discharging rotating valve. — p = Wheel for reversing rotating valve. — q = Wheel for engaging the friction coupling. — r = control table.— Öl = Oil. — Wasser = Water.

interrupting the cooling when long stops are made at stations or when running downhill. For the circulation of the cooling water for the liquid transmission a piston pump is provided which is driven by the main engine.

A small auxiliary compressor, which may be driven by the auxiliary engine, allows of the engine being started up under any conditions.

For both travelling directions there is a control board with the necessary handles etc.

Fig. 50. Schwartzkopff Diesel locomotive with hydraulic transmission.

The following are the main dimensions of the locomotive:

Gauge . . . . 1435 mm (4′ 8½″)
Dia. of driving wheel . . . . 1250 mm (4′ 1¼″)
Dia. of crank circle . . . . 500 mm (19³/₄″)
Max. power of engine . . . . 220 I.H.P.
Revs./min of engine shaft 450
Max. tractive force at draw-hook . . . . 5000 kg (11000 lbs.)
Tare . . . . 42 tons
Weight in working order . . . . 44 ,,
Max. speed . . . . 40 km/h (25 m.p.h.)
Capacity of cooling water tank . . . . 1500 litres (330 glns.)
   ,,   ,, fuel tank 600   ,,   (130 ,,)

Trial runs were made on lines of the German State Railway with trains of 9 to 10 coaches with a total weight up to 150 tons (see fig. 51), with the following results:

Fig. 51. Train hauled by Schwartzkopff Diesel locomotive with hydraulic transmission.

Efficiency of driving gear 76%
Fuel consumption per hour (Roumanian gasoil). . . . . 39.2 kg (86 lbs.)

Fuel consumption { at the drawhook . . . . 350 g (0.77 lb.)
 per H.P. hour { at the wheel diameter . . . . 270 g (0.60 lb.)
{ at the engine shaft . . . . 187 g (0.41 lb.)

2 C 2 (4—6—4) DIESEL LOCOMOTIVE
FOR THE GERMAN STATE RAILWAY.

This locomotive, shown in fig. 52, was built by the Esslingen Engineering Works in cooperation with the Augsburg-Nürnberg Engineering Works, which latter firm built the Diesel engine.

The transmission is with compressed air. The engine is a four-stroke Diesel engine of 1000/1200 H.P., of the same type as used for the Russian 1 E 1 (2—10—2) locomotive but compressorless. A two-cylinder double acting air compressor is coupled direct to the Diesel engine and the compressed air is heated by the exhaust gas from the engine in counter-current. The machinery is on the same principle as that of an ordinary steam locomotive, except that hot air is used instead of steam. In design and working, therefore, the whole is similar to a steam engine, the only difference being that the boiler is replaced by an installation for delivering compressed air. The peculiar feature of this system is that in the hot air there is more potential energy than is delivered by the Diesel engine to the shaft. Thus, by utilizing the heat contained in the exhaust gas of the Diesel engine, a total efficiency of 26% can be reached, calculated from fuel to draw-hook.

The main dimensions of this locomotive are:

		Continuous:	*Short time:*
Engine power E.H.P.		1000	1200
„ „ I.H.P.		1350	1630
Revs. of engine shaft per min		400	450
Working pressure of air { in atm.		6.5	7
{ in lbs./sq. in.		92	100

Air temp. at cylinder inlet . . . . 320° C. (608° F.)

Tractive force at driving wheel circ. . . . . 11200 kg (24500 lbs)

Diesel { diameter . . . . 450 mm (17³/₄″)
cylinders { stroke . . . . 420 mm (16½″)
{ number 6

Compressor { diameter . . . . 640 mm (25¹/₄″)
cylinders { stroke . . . . 350 mm (13³/₄″)
{ number . . . . 2 (double action)

Locomotive { diameter . . . . 710 mm (28″)
cylinders { stroke . . . . 650 mm (25⁵/₈″)
{ number . . . . 2 (double action)

Dia. of driving wheels . . . . 1600 mm (5′ 3″)
Rigid wheel base . . . . 4.70 m (15′ 5″)
Distance centre to centre of bogies . . . . 10.30 „ (33′ 10″)
Total wheel base . . . . 12.50 „ (41′)

Fig. 52. Diesel compressed air locomotive of the German State Railway Company.

Length over buffers . . . . 15.80 m (51' 10")
Tare . . . . 111 tons
Weight in working order . . . . 120 „
Adhesion weight . . . . 54 „
Max. speed . . . . 80 km/h (50 m.p.h.)

In order to minimize the clearance in the compressor only 1 mm distance has been left between the piston and the cylinder covers at the dead centres, thus reducing the clearance to 2.4%.

Further, in order to prevent the temperature rising too high in the compressor it was found necessary to provide a very effective cooling, which could not very well be attained with cooling jacket and piston cooling alone, so that instead of this a system of spray cooling has been applied, which consists in cooling water being sprayed into the cylinders in mist form by means of a spray pump minutely adjustable in order to keep the temperature constant at a certain level not exceeding 200° C. at the outlet. It appears, however, that the compressor can safely be run for a time without cooling (though, of course, with diminishing efficiency), which has the advantage that in the event of a breakdown in the cooling system while the locomotive is running it can still continue for another 30 to 45 minutes.

The water must be as pure as possible in order to prevent the formation of scale in the compressor cylinders, and to this end a jacket is placed around the exhaust pipe of the engine, through which the cooling water is first passed and thus heated to 100—120° C., thereby depositing a large part of the scale. The data given below in regard to the compressor give some idea of the difficulties which had to be surmounted in the design in consequence of the large volume of air that has to be compressed and the temperatures thereby arising:

Max. quantity of air to be compressed: about 200 kg (440 lbs.) per min.
Spraying water required: about 6 kg (13.2 lbs.) per minute.
Temp. of compressed air: about 200° C. (392° F.).

Before being delivered to the locomotive cylinders, the compressed air, which is somewhat saturated with water vapour in consequence of the spraying, is heated again to a maximum temperature of 350° C. (662° F.) in a heater with 82.5 sq. metres (890 sq. ft.) heating surface and consisting of several nests of tubes. The air passes along the outside of these tubes, whilst the exhaust gas is led through them, the respective rates of flow being about 15 and 45 metres (49 and 148 ft.) per second.

In this locomotive there are three different types of coolers (mounted on the end partitions), one for cooling the cylinder cooling water, one for the piston cooling oil and another for the spray-cooling water. Electrically-driven ventilators provide for a continuous flow of air.

The locomotive cylinders are fitted with ordinary piston valves but with a double charge, in view of the large quantities of air which have to be supplied. The degree of expansion is regulated in the same way as with a steam locomotive, whilst a governor is similarly provided for throttling the air supply from the heater to the locomotive cylinders. In addition to the link lever and the throttle valve the driver has also to attend to the fuel and speed regulators and the spray pump for the compressor cooling water.

A long trial run was made on 22nd November 1929 as an express train with a weight of 233 tons from Stuttgart to Augsburg, on which journey the "Geislinger Stiege", a gradient of 1 : 40 about 6 kilometres long, was taken at a speed of 20 kilometres (12 miles) per hour without the aid of a bank engine behind and without the Diesel engine being fully loaded.

This first long trial run fully demonstrated the advantages of pneumatic transmission. It was found that even under difficult conditions, for instance in sharp curves on a gradient, the locomotive could easily be started and that speed could be got up much quicker than is the case with a steam locomotive of equal power. Further, the capacity of the coolers was found to be quite adequate and on the trial run mentioned it was not necessary to use the ventilators for cooling.

A two-days' run was made by the locomotive on its own power from Esslingen to the locomotive testing plant at Grunewald and, notwithstanding the fact that the maximum speed of 80 km (50 miles) per hour was maintained the whole of the way, no troubles whatever were experienced.

DIESEL LOCOMOTIVE OF THE "MASCHINENFABRIK ESSLINGEN".

This locomotive has a power of 150/165 H.P. and is normally constructed with gear wheel transmission. The Diesel engine is mounted crosswise, so that all shafts are parallel to the driving axles of the locomotive, affording a simple transmission arrangement.

Fig. 53. Diesel locomotive with gear wheel transmission
of the "Maschinenfabrik Esslingen".

The engine itself is a single-acting, compressorless, four-stroke MAN Diesel engine.

The number of revolutions of the engine is regulated by hand by adjusting the fuel pumps. Furthermore, the moment of fuel injection can be advanced or retarded so as to get the optimum conditions for injection and combustion with varying speed and load. A governor is also provided to avoid an excessive number of revolutions. The engine is started by a *Bosch* dynamo charged from a battery, and in order to facilitate starting a decompression device is installed by means of

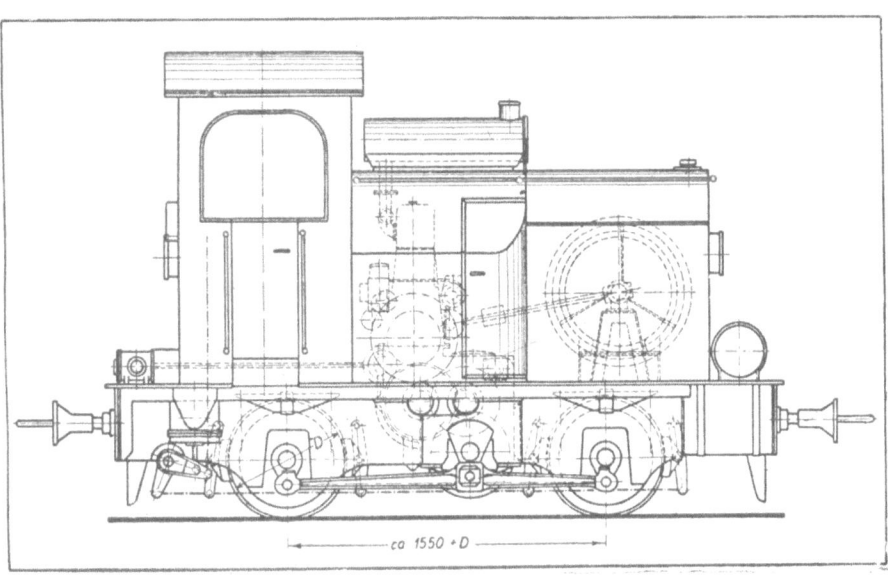

Geschwindigkeiten V km/Std. u. Zugkräfte am Radumfang Zr kg bei Raddurchmesser D mm								
	I	II	III	IV	V	VI	VII	D
V	6,7÷31,4 (35)	7,2÷34 (37,5)	7,8÷36,5 (40,5)	8,4÷39 (43,5)	9,0÷42,5 (47)	9,6÷45 (50)	10,3÷48 (53,5)	700
Zr	4600÷990	4250÷915	3950÷850	3700÷785	3440÷730	3220÷690	3000÷645	
V	7,2÷34 (37,5)	7,7÷36 (40)	8,4÷39 (43)	9,0÷42 (46,5)	9,6÷45,5 (50)	10,3÷48 (53,5)	11,0÷51,5 (57)	750
Zr	4300÷920	4000÷860	3700÷795	3440÷735	3200÷680	3000÷645	2800÷600	
V	7,7÷36 (40)	8,3÷39 (43)	8,9÷42 (46)	9,6÷45 (50)	10,3÷48 (53,9)	11,0÷51,5 (57)	11,8÷55 (61)	800
Zr	4000÷860	3720÷795	3480÷735	3200÷690	3000÷645	2800÷600	2620÷560	
V	8,1÷38 (42,5)	8,8÷41 (46,5)	9,5÷44 (49)	10,2÷47,5 (53)	10,9÷51,5 (57)	11,6÷55 (61)	12,5÷58 (65)	850
Zr	3820÷815	3520÷755	3260÷705	3040÷650	2840÷600	2660÷565	2480÷535	
V	8,6÷41 (45)	9,3÷43,5 (48,5)	10,0÷47 (52)	10,6÷50 (56)	11,6÷55 (60)	12,4÷58 (64)	13,2÷62 (69)	900
Zr	3600÷755	3340÷710	3100÷660	2860÷620	2660÷560	2500÷535	2340÷500	
V	9,1÷42,5 (47)	9,8÷46 (51)	10,3÷49,5 (55)	11,4÷53 (59)	12,2÷58 (64)	13,0÷61 (68)	14,0÷65 (73)	950
Zr	3400÷730	3160÷675	3000÷625	2700÷585	2540÷535	2380÷505	2200÷475	
V	9,6÷45 (50)	10,3÷48,5 (54)	11,1÷52 (58)	12,0÷56 (62)	12,8÷61 (67)	13,7÷64 (71)	14,7÷69 (77)	1000
Zr	3220÷690	3000÷640	2800÷595	2580÷550	2420÷505	2260÷485	2100÷450	

Fig. 54. Diesel locomotive with gear wheel transmission
of the "Maschinenfabrik Esslingen".
One may choose between the values I—VII of speeds V and tractive forces at driving
wheel circumference Zr for a given wheel diameter D.
Geschwindigkeiten = Speeds. — Zugkräfte am Radumfang = Tractive forces at
driving wheel circumference. — bei Raddurchmesser = at wheel diameter.

which for the first strokes the air suction valve is kept open, so that the suction air is not compressed.

The dimensions of the cylinders are:

Diameter . . . . 165 mm (6½")
Stroke . . . . 220 „ (8⅝")
Continuous power at 900 revs./min . . . . .150 H.P.
Temporary max. power at 1000 revs./min . . . . .165 H.P.

Transmission is effected with spur gear wheels, which are continuously in mesh. The gear wheels are coupled onto the shaft or released by means of hydraulic multiple disc couplings. Four-speed or three-speed couplings and two direction couplings are provided.

This locomotive is illustrated in figs. 53 and 54.

SHUNTING LOCOMOTIVE OF THE SWISS FEDERAL RAILWAYS.

This petrol locomotive, built by the "Schweizerische Lokomotiv- und Maschinenfabrik", Winterthur, is doing service as shunting locomotive on the wharves and for the grain silos at Lucerne.

The main dimensions are:

Diameter of wheel . . . . 850 mm (2' 9½")
Wheel base . . . . 2.60 m (8' 6½")
Length overall . . . . 6.54 m (21' 5½")

Engine ⎰ power . . . . 100 H.P.
        revs./min 1200
        cylinders 6
        bore . . . . 150 mm (5⅞")
        stroke . . . . 170 mm (6¾")

Speeds . . . . 3.75; 7.5; 15 km/h (2.3; 4.6; 9.3 m.p.h.)
Max. tractive force . . . . 5500 kg (12000 lbs.)
Weight in working order . . . . 21.3 tons

Fig. 55. Shunting locomotive Swiss Federal Railways.

The frame consists of two separate parts, the frame proper carrying the two driving axles, engine and transmission, and another frame bearing the driver's cabin and resting on the main frame in three places.

The motion of the engine shaft is transmitted by a flexible clutch to the speed gears with oil pressure control and thence via the reversing gear to the driving axle, which latter is connected to the coupled axle by coupling rods.

The upper frame carrying the driver's cabin is sprung on the lower frame and can easily be removed to give easy access to the engine and gearing.

The engine is provided with a flywheel, a ventilator, an electric (starting) motor with gear wheel transmission on the flywheel, an electric light motor with accumulator battery, a carburettor, etc.

The speed gearing with oil pressure control and the reversing gear are combined in one box. The whole gearing runs in an oil bath.

On the level the locomotive can haul a train load of 250 tons at a speed of 15 km/h (9.3 m.p.h.) and can start with a train load of 350 tons. Up a gradient of 1 : 100 the hauling power is 170 tons at a speed of 7½ km/h (4.6 m.p.h.).

A similar locomotive of the same nominal power supplied to a private firm has set a train load of 53 tons in motion on a gradient of 1 : 18 in a curve. This locomotive makes 5 or 6 runs every day from the works to the station with a train load of 30—50 tons. The radius of the sharpest curve on the factory site is 45 metres (50 yards), which curve lies on a gradient of 1 : 33.

The locomotive described is illustrated in fig. 55.

WINTERTHUR DIESEL LOCOMOTIVE FOR SIAM.

This locomotive (see fig. 56) was built by the "Schweizerische Lokomotiv- und Maschinenfabrik" at Winterthur (Switzerland) for light traffic on the Siamese State Railways.

Fig. 56. Diesel locomotive Siamese State Railways.

6

The engine is a six-cylinder, compressorless, four-stroke Diesel engine; the small compressor built onto the engine is only for compressing air for starting. The transmission is hydraulic, with five speeds, on the S L M Winterthur system, with reversing mechanism acting on the loose axle; both parts are mounted in a cast-steel box.

Two multiple plate tube coolers are provided for cooling the engine with water, while air is drawn in through the cooling elements by ventilators driven by the engine shaft, the air then passing over the engine and led off underneath.

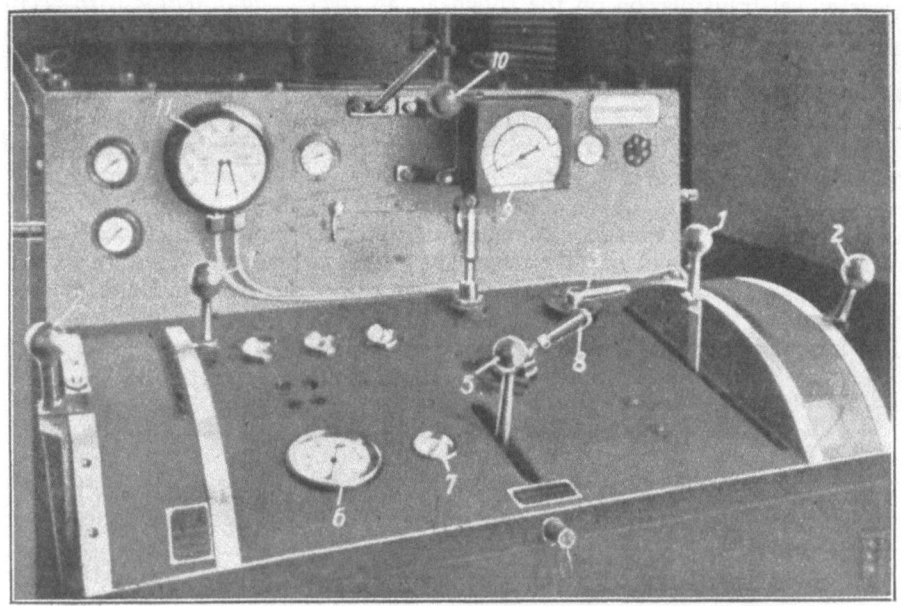

Fig. 57. Control board Diesel locomotive Siamese State Railways.

1. Fuel regulator.
2. Lever for reversing gear.
3. Fuel valve.
4. Auxiliary valve for compressed air vessel.
5. Reversing gear.
6. Tachometer.
7. Cooling water distance thermometer.
8. Sanding lever.
9. Speed gauge.
10. Whistle.
11. Brake vacuum meter.

An ordinary pipe cooler is also provided for each multiple plate tube cooler for cooling the engine and the lubricating oil of the driving machinery.

The reversing gear, the sand box and the whistle are worked with air of 7 at (100 lbs.) drawn from the starting-air vessel via a reducing valve.

The locomotive is designed for one-man operation; the control board is shown in fig. 57.

Trial runs were carried out with one of these locomotives in 1928 on the Land-quart-Chur-Disentis line, which is 74 kilometres (46 miles) long. Particulars concerning these trials are given in fig. 58, from which it is seen that the steepest gradient is 27 : 1000, whilst there are many curves with a radius up to 120 metres (130 yards), particularly above Reichenau. The train left Landquart with a total weight

of 70 tons, of which 23.5 tons for the locomotive itself. At Chur, Felsberg, Reichenau, Versam, Ilanz, Tavanasa and Truns more or less long stops were made, during
which time some shunting was done, which afforded an opportunity to demonstrate
the good points of the SLM transmission and the simple operation. As far as Ta-

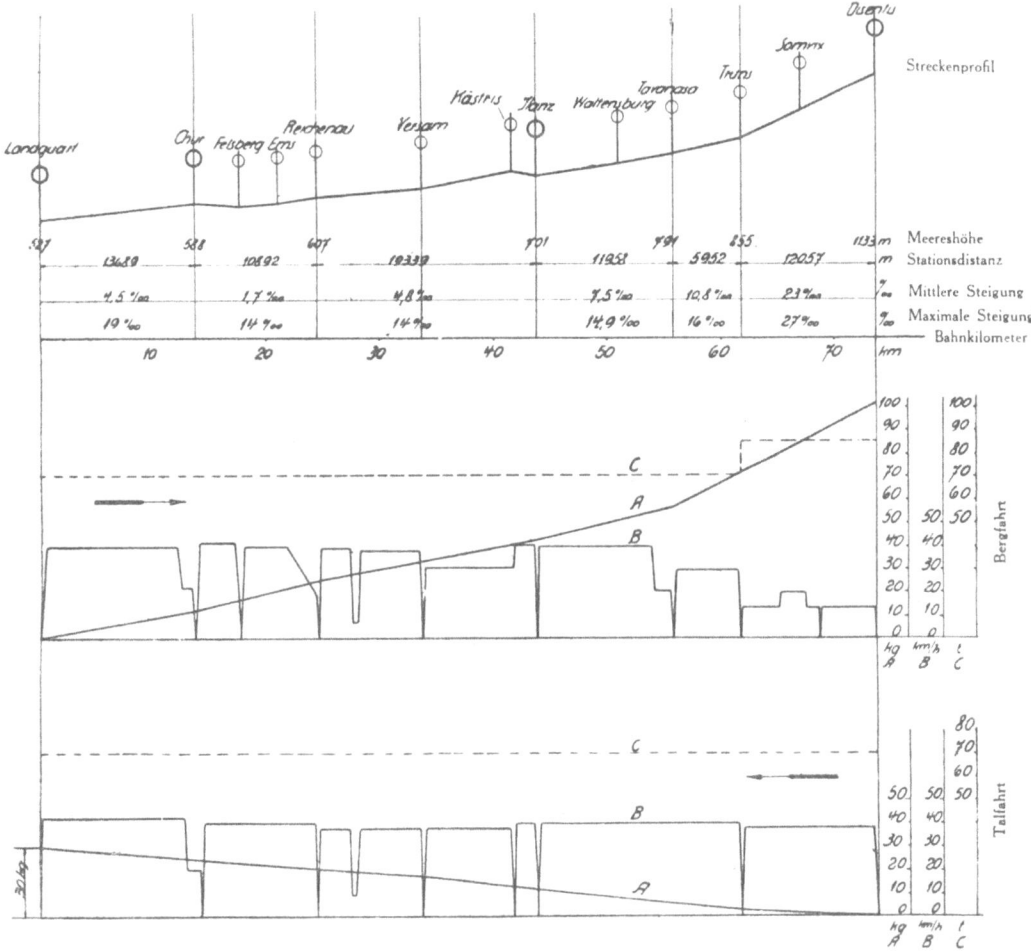

Fig. 58. Diagram of trial run.
A = Fuel consumption. — B = Speed of train. — C = Weight of train.
Bahnkilometer = Kilometer posts. — Bergfahrt = Mountain run. — Maximale Steigung = Max.
gradient. — Meereshöhe = Sea level. — Mittlere Steigung = Average gradient. — Streckenprofil =
Profile of the line. — Stationsdistanz = Distance between stations = Talfahrt = Valley run.

vanasa the train ran in fifth gear at an average speed of 40 km/h. Between
Tavanasa and Truns the fourth gear was taken at 28 km/h and the last 12 kilometres from Truns to Disentis was covered at a speed of 14 km/h in second gear,
with a total train weight of 85 tons. A starting test on the gradient of 27 : 1000 with
a train weight of 85 tons was satisfactory. The return journey from Disentis to
Landquart took exactly 2½ hours, delay due to stopping included.

The fuel used was a gasoil of 0.87 specific gravity with a minimum calorific value of 10030 kcal/kg (18000 B.T.U. per lb.).

The consumption on the above trial runs was:
on level stretch Landquart-Ilanz:

out	30	kg	=	66 lbs.
back	18.5	,,	=	41 ,,
total	48.5	kg	=	107 lbs.

which works out at 7.9 grammes per ton per kilometre (0.028 lb./ton/mile);
on the hilly stretch Ilanz-Disentis:

out	71.5	kg	=	155 lbs.
back	11.5	,,	=	25 ,,
total	83	kg	=	180 lbs.

which is 19 g/tkm (0.067 lb./ton/mile).

These satisfactory results were due partly to the low fuel consumption and partly to the high efficiency of the transmission.

The results with the engine and transmission on the test bed, before these were built in the locomotive, were as follows:

fuel consumption of engine:

179	g	(0.395 lb.)	per E.H.P./h at	180	E.H.P.
184	,,	(0.405 ,,)	,,	,,	,,	,,	150	,,
205	,,	(0.45 ,,)	,,	,,	,,	,,	100	,,

efficiency of transmission, depending on the load:

in 2nd gear	95—97%
in 3rd ,,	93—95%
in 4th ,,	92—94%
in 5th ,,	90—91%

1 C 1 (2—6—2) DIESEL LOCOMOTIVE FOR THE ITALIAN STATE RAILWAYS.

This locomotive (see fig. 59) was built by the "Maschinenbau Gesellschaft" of Karlsruhe for the Italian State Railways in Sardinia. The Diesel engine is a compressorless four-stroke Benz engine; the transmission is on the Lentz hydraulic system; the locomotive is fitted with electric lighting, compressed air brakes and sand-box operated by air.

The main dimensions of the locomotive are:

Normal engine power	200 H.P.
Max. engine power	250 ,,
Gauge	950 mm (3′ 1/3/8″)
Max. speed	37 km/h (23 m.p.h.)
Weight in working order	about 40 tons
Max. axle load	8.4 tons
Max. starting tractive power at drawhook	4000 kg (8800 lbs.)
Average fuel consumption per running hour about	28 kg (65 lbs.)

Fig. 59. Diesel locomotive of the Italian State Railways.

Similar locomotives built by the same firm are in service on the Cyrenaica State Railways in the Italian dependency of Libya (North Africa).

ZARLATTI DIESEL LOCOMOTIVE.

A trial has been made in Italy with a Diesel locomotive built on the Zarlatti system. This Zarlatti locomotive, reconstructed from a steam locomotive class 910 of the Italian State Railways, has a Diesel-Fiat S-Giorgio engine which drives a two-phase rotatory compressor at a speed of 480 revs./min. The compressor, supplied by the "Lokomotiv- und Maschinenfabrik Winterthur", delivers air under a pressure of 142 lbs./sq. inch (10 kg/cm^2) which, after being saturated with steam produced in an oil-fired auxiliary boiler (Wickers-Terni), enters the locomotive cylinders. The heat of the engine exhaust gas is used for preheating the boiler feed water (via an economiser).

The inventor claims that the efficiency of the whole aggregate — Diesel engine, air compressor, boiler and cylinders — is much higher than that of a normal steam locomotive (boiler and cylinders), due not only to the higher efficiency of the Diesel engine alone but also to the fact of the heat stored in the exhaust gas and in the returned cooling water being utilised for preheating the boiler· feed water, whilst also the boiler itself has a higher efficiency and, finally, better use is made of the expansive force of the mixture of the air and steam as compared with steam alone. In consequence, it is claimed, the locomotive acquires that valuable flexibility of power characteristic of steam locomotives.

Experience has yet to show in how far these advantages are actually realised and whether they are not outweighed by the complications of the Zarlatti locomotive (boiler, combustion engine and compressor). One advantage that the Zarlatti system certainly has is that in reconstruction some parts of the existing locomotives can be used (cylinders with steam-distributing mechanism, frame, etc.).

Trials carried out with this locomotive on the Benevento-Cancello line in October 1927 yielded very satisfactory results as regards both running and power. The first five trial runs made under normal load showed that the ratio of the oil consumption for Diesel engine and boiler to the weight of coal required for the same amount of work was as 1 : 6, which means a cash saving of more than 50%, taking coal at 185 lire per ton and oil at 440 lire per ton. The maximum speed reached during these runs was 50 km/h (31 m.p.h.).

Since these trials the Zarlatti locomotive has apparently never been taken into regular service on the Benevento-Cancello line.

ANSALDO DIESEL LOCOMOTIVE.

This locomotive, built by the Ansaldo S.A. at Genova-Sampierdarena, has a direct drive and is started with the aid of compressed air.

The engine is of the two-stroke type and has six horizontal cylinders placed in two rows of three one above the other. Each cylinder has two pistons with a joint combustion chamber, the pistons thus working in opposite directions. The piston rods on the same side of each pair of cylinders are connected up by an oscillating lever and each lever is connected by rods to the driving wheels by means of a crankshaft placed crosswise in the frame of the locomotive and above the group of cylinders.

This arrangement provides a simple transmission, whilst the installation of a special motor allows of a simple construction in which cylinder covers are dispensed with, as also the inlet and outlet valves, the functions of which are performed by the two pistons in opening and closing the ports (one for combustion gas and the other for fuel and scavenging air) towards the end of the forward stroke and at the beginning of the return stroke; at the same time a good scavenging action is obtained through the air current passing in only one direction.

The weight is reduced to a minimum by the absence of a special motor frame, for this is formed by the locomotive frame itself. This lessens the possibility of damage, as all the complicated and delicate machinery for the distribution is no longer necessary, the distribution being effected by special oil atomizers which allow of a variation of the fuel feed corresponding to changes in the road. Starting is done in the ordinary way.

The starting of the train is done independently of the engine by means of compressed air delivered to the scavenging pump cylinders, which are mounted on the outside of the locomotive frame in the same way as cylinders of a steam locomotive and act in a similar manner until the train has got up sufficient speed for the Diesel engine to be set going, which is usually the case at 10 revs./min. The Diesel engine then supplies the necessary power alone and the scavenging pumps assume their proper function by drawing in air and delivering it to the cylinders of the Diesel engine.

Thus the locomotive may be said to have two separate engines, one for starting with compressed air and a Diesel engine which begins to work when the train has got up a certain speed, without, however, involving any appreciable increase in weight or further complications, in view of the air engine being formed from parts which are in any case required for the Diesel engine.

Reversing is effected in a simple manner with the air engine alone by means of an arrangement exactly similar to the steam-distributing mechanism in a steam locomotive. Once the locomotive has been set in motion by the air engine the Diesel engine propels the train farther in the same direction.

Trials have shown the fuel consumption to be 7—8.5 g/tkm (0.025—0.03 lb./ton mile), and in some cases 10 g/tkm (0.035 lb./ton mile). Against a consumption of 0.043—0.067 kg (0.153—0.235 lb.) coal for an ordinary steam locomotive, the Diesel oil consumption of this locomotive is 0.0062—0.0085 kg (0.0215—0.03 lb.), both per ton (excl. locomotive) and per kilometre (or per ton mile). The cost of fuel per running tkm at a coal price of 140 lira and an oil price of 450 lire per ton is 0.0077 lira for a steam locomotive and 0.0033 lira for the Diesel locomotive.

The main dimensions of the Ansaldo Diesel locomotive are:

Gauge . . . . 1445 mm (4'8$^7/_8$")
Type of engine: two cycle horizontal; number of cylinders 6
Dia. of cylinders . . . . 330 mm (13")
Stroke . . . . 2 × 480 mm (18$^7/_8$")
Max. revs./min 300
Power at engine shaft . . . . 1100 H.P.
Dia. of driving wheels . . . . 1370 mm (4'6")
Dia. of bogie wheels. . . . . 980 ,, (3' 2½")
Dia. of Bissell bogie wheels . . . . 1110 ,, (3' 7$^3/_4$")
Rigid wheel base . . . . 4100 ,, (13' 5½")
Total wheel base . . . . 10750 ,, (35' 3$^1/_4$")
Total length over buffers . . . . 14200 ,, (46' 7")
Adhesion weight . . . . 45 tons
Total weight in working order . . . . 84 ,,
Total tare . . . . 80.5 ,,
Max. speed (300 revs. of engine shaft) . . . . 75 km/h (46 m.p.h.)
Max. power on driving wheel dia. at 65 km/h (40 m.p.h.) 935 H.P.
Corresponding tractive force . . . . 4200 kg (9250 lbs.)
Power at driving wheel circ. at 10 km/h (6.2 m.p.h.)
    with 40 revs./min of engine . . . . 210 H.P.
Corresponding tractive force . . . . 5600 kg (12300 lbs.)
Capacity of oil tank . . . . 1200 kg (2650 lbs.)
Oil consumption per E.H.P.h . . . . 0.2 kg (0.44 lb.)

KITSON-STILL LOCOMOTIVE.

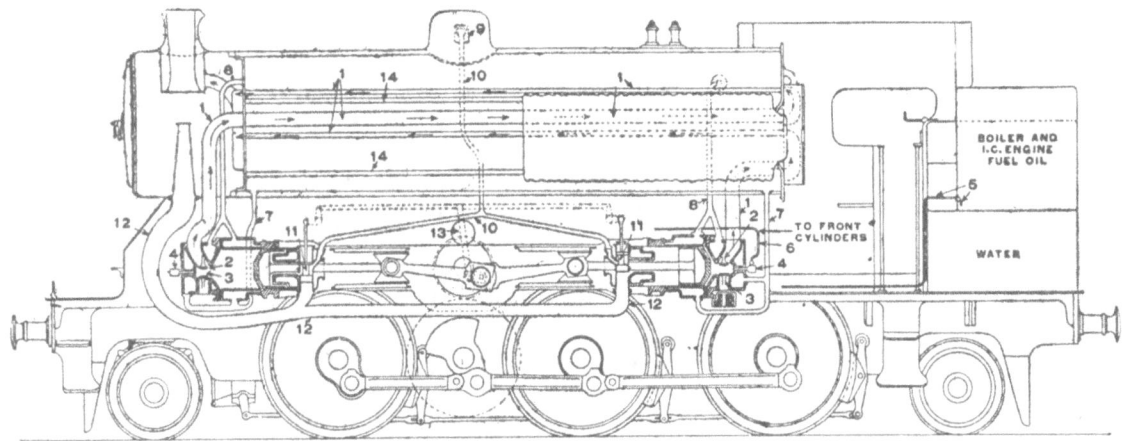

Fig. 60. Kitson-Still Locomotive.

The Kitson-Still locomotive is a 1 C 1 (2—6—2) type locomotive built by Kitson & Co. Ltd. of Leeds for ordinary main line service. The engine has double action, as combustion engine on one side of the piston and as steam engine on the other side. The cooling water in the cylinder jacket is in direct connection with the boiler. The heat of the engine exhaust gas is used for generating steam in the boiler. As shown in fig. 60, there are three coupled axles.

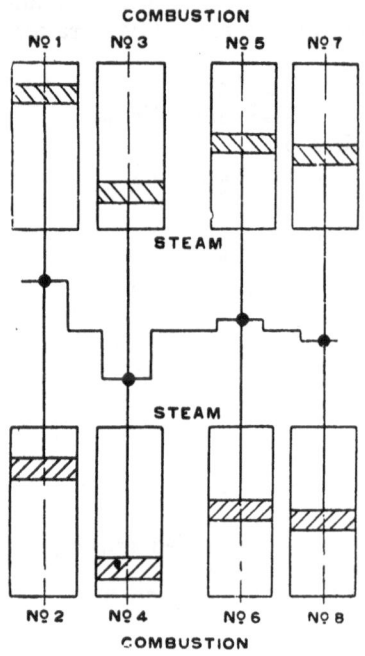

Fig. 61. Cylinder diagram of the Kitson-Still locomotive.

A four-stroke combustion engine was chosen because with a two-stroke system the cylinders would have been too small for starting with steam alone, unless the combustion side of the pistons were also brought into action as a steam engine, which would have led to undesirable complications.

The cylinders are mounted horizontal, because for vertical cylinders short connecting rods would have to be used, as otherwise the aggregate would be too high and, moreover, the transmission to the driving axles would be more complicated. On the other hand it was considered better not to have the cylinders acting direct on the driving axles, because in that case owing to the play in the springs the cylinders could not be kept at a constant distance from the driving axles. For these reasons the cylinders are arranged so as to work on a loose crank shaft fixed in the locomotive frame. The movement of this crank shaft is transmitted to a loose axle connected by coupling rods to the driving axles.

With a view to maintaining a positive driving torque eight cylinders are provided, mounted in two rows of four on either side of the crank shaft (see fig. 61), whilst in order to avoid difficulties with the stuffing boxes all cylinders have a combustion action on the cover side and a steam action on the shaft side.

The following are the main dimensions of the locomotive:

Cylinders:	number	8
	stroke	15½" (394 mm)
	bore	13½" (343 mm)
Gear wheel transmission ratio		1.878 : 1
Max. revs. engine shaft		450 per min
Max. speed of locomotive		45 m.p.h. (72 km/h)
Max. power	combustion engine alone	1000 I.H.P.
	total available	1200 I.H.P.
Starting tractive power		25450 lbs. (11500 kg)
Tractive power at max. speed		7000 „ (3200 „)
Boiler (oil fired)	number of tubes	119
	ext. dia. of tubes	1¾" (45 mm)
	length of tubes	9' (2.743 m)
	heating surface of tubes	491 sq. ft. (46 m²)
	„ „ „ firebox	72 „ „ (7 m²)
	„ „ „ total	563 „ „ (53 m²)

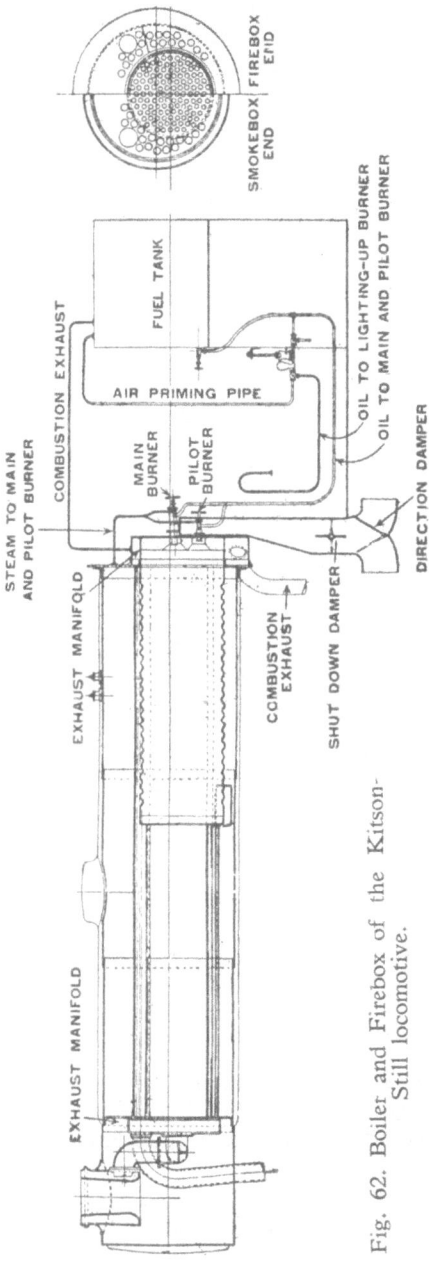

Fig. 62. Boiler and Firebox of the Kitson-Still locomotive.

Regenerator (heated with exhaust gas)	number of tubes 36 + 2
	ext. dia. of tubes $2^3/_4$" (70mm) & $6^1/_2$" (165 mm)
	length of tubes . 17' 4" (5.283 m)
	heating surface 508 sq.ft. (47 m²)

Storage	water . . 1000 glns. (4500 litres)
	fuel oil . . 400 „ (1800 „)
	lubricating oil 85 „ (40 „)

Weights in working order	on each pilot or trailing axle 9.5 tons
	„ „ coupled axle . . 17 „
	adhesion . . . . 51 „
	total . . . . 70 „

The engine is reversible. All parts are lubricated under pressure. The boiler (fig. 62) is of simple construction and comprises an inner boiler serving as firebox and extending about half the length of the outer boiler. Joining on to this inner boiler is a nest of flue tubes extending to the smoke-box. On both sides of the inner boiler and the nest of flue tubes is another nest of tubes extending the whole length of the boiler and acting as regenerator; through these tubes the exhaust gases are passed from the combustion engine. For starting up, the boiler is oil-fired in the fire-box. The steam-raising capacity of the boiler is 8000 lbs. (3600 kg) per hour.

In the driver's cabin there are handles for the steam regulator, oil regulator, reversing gear, brakes, cylinder cocks, sand-box, air valves, etc.

The locomotive is set in motion by the following operations: the reversing lever is moved to "full ahead" (or "reverse"), whereby the valve mechanism of the combustion engine is set in the right position at the same time; the oil regulator is moved to the starting position, the steam regulator opened and the locomotive begins to move. As soon as the engine has made a few revolutions the combustion engine begins to work on low load, and as the engine gains speed the reversing lever is drawn up and the oil regulator opened wider. At a speed exceeding 6 m.p.h. (10 km/h) the steam regulator can be closed entirely, only the combustion engine then being used.

To stop the locomotive the oil regulator is closed (also the steam regulator if this should be open) and the brakes applied in the usual way. In emergencies counter steam may be given by throwing over the reversing lever to the reversed running position and opening the steam regulator.

For starting up under unfavourable conditions, e.g. with a heavy train on a gradient or in a sharp curve, the compression in the combustion engine can be switched off for a few revolutions by placing the valve mechanism for the combustion engine in the centre position, thereby lifting all the valves and cutting off the fuel charge.

Steam is used not only for starting but also for the vacuum brakes and the heating of the train.

Tests have shown that the maximun fuel consumption is 0.37 lb. (168 g) per I.H.P., whilst the water consumption is one-tenth of that of a steam locomotive of the same power.

The locomotive described was tried out on the London and North Eastern Railway in the neighbourhood of Leeds, with a train consisting of a six-wheeled ob-

Fig. 63. Kitson-Stil Locomotive.

servation car and six coaches coupled in pairs as double coaches on three bogies. The total weight of the train (excl. locomotive) was 118 tons (see fig. 63).

On the lines over which the trials took place there are many gradients, the steepest of which is 1 : 50.

The trials proved that the locomotive is suitable for hauling trains like the one described and under varying conditions, i.e. for long non-stop runs as well as for runs with frequent stops.

DIESEL-ELECTRIC LOCOMOTIVE OF THE LONDON AND NORTH EASTERN RAILWAY.

This is a rebuilt electric locomotive used on the electrified line Newport-Shildon. It rests on two double-axled bogies and all axles are driven.

The Diesel engine, Beardmore make, rated at 1000 H.P., has variable speed and is coupled to a direct current generator of 800 volts, with separate exciting dynamo. This generator supplies the power for the four traction motors, which drive the axles with a transmission ratio of 1:4.5.

Buffers and draw gear are built on the two bogies, so that the body of the locomotive is free of shocks or pulling forces.

There is a driver's cabin at both ends of the locomotive.

Low tension current is supplied by an auxiliary dynamo and battery to an electrically driven air compressor for the continuous air-brake.

No provision has been made for heating the train.

The main dimensions of the locomotive are:

Total weight . . . . 90 tons
Adhesion weight . . . . 90 „
Max. tractive force . . . . 18 „
Continuous power at 43½ km/h (27 m.p.h.) . . . . 775 H.P.
Distance between bogies (centre to centre) . . . . 18′ 3″ (5.563 m)
Total wheel base . . . . 27′ 0″ (8.230 m)
Length over buffers . . . . 39′ 4″ (11.939 m)

This locomotive is destined for goods traffic for which at present 1 D (2—8—0) steam locomotives are used.

Fig. 64. 60 H.P. Kerr Stuart Diesel Locomotive.

C (0—6—0) NARROW GAUGE DIESEL LOCOMOTIVE
OF KERR STUART & CO., LTD.

At the California Works of Kerr Stuart & Co., Ltd at Stoke-on-Trent a narrow gauge C (0—6—0) Diesel locomotive (fig. 64) has been built and tried out on the Welsh Highland Railway. The Diesel engine is of the four-stroke type made by J. & H. McLaren Ltd, Leeds. The fuel is pumped into a fore-chamber provided on top of each cylinder.

Starting is done with a 4 H.P. petrol motor and takes 2½—3 minutes.

Transmission is by means of gear wheels and chains.

The main dimensions are as follows:

Gauge . . . . 1' 11½" (597 mm)
Weight in working order . . . . 10 tons
Engine { cylinders 4
bore . . . . 135 mm ($5^5/_{16}$")
stroke . . . . 200 „ ($7^7/_8$")
power . . . . 60 H.P.

The data for the hauling power of this locomotive are tabulated below:

	Speed		Trailing load in tons	
	m.p.h.	km/h	level track	gradient 1 : 20
2nd speed	11.27	18	133	7
	8	13	150	9
1st speed	5.8	9.2	268	22
	4	6.5	302	27

The same firm builds similar locomotives with two and three axles for 30 to 90 H.P. and for gauges of 2 to 3½ ft. (600—1067 mm).

AVONSIDE LOCOMOTIVES.

The Avonside Engine Co. at Bristol build locomotives of 30 to 120 H.P. with two and three driving axles, one of which is driven by gear wheel transmission and chain, the transmission to the other coupled axles being effected by means of coupling rods.

The small locomotives are made for two speeds, the medium sized ones for two or three, and the large ones for three or four speeds.

Further they are made for normal gauge 4'8½" (1.435 m), for narrow and for broad gauges.

Such a locomotive is illustrated in fig. 65, the main dimensions of which are the following:

Engine power . . . . 80—100 H.P.
Speeds . . . . 4, 8 and 16 m.p.h. (6½, 13 and 26 km/h)
Tractive force { Speed I . . . . 5300—6600 lbs. (2400—3000 kg)
„ II . . . . 2600—3300 „ (1200—1500 kg)
„ III . . . . 1300—1600 „ (600— 700 kg)

Fig. 65. Avonside Locomotive.

The lowest powers apply to locomotives with Diesel engines, the highest to locomotives with petrol-engines. The engines are of the four-cylinder, four-stroke type. The Diesel engines have airless injection.

LOCOMOTOR OF THE MOTOR RAIL AND TRAMCAR CO., LTD, BEDFORD.

Fig. 66. Petrol Locomotor of the Motor Rail & Tram Car Co. Ltd.

This locomotor is illustrated in fig. 66 and is used on several English railways, among others on the London and North Eastern Railway.

The engine is a four-cycle petrol engine and the transmission is for two speeds in either direction. The locomotor is provided with sand-boxes and an exhaust whistle on the silencer.

The main dimensions are:

Engine power 40 H.P.
Number of
  cylinders . 4
Cylinder bore 120 mm ($4\frac{3}{4}''$)
Stroke . . . 140 mm ($5\frac{1}{2}''$)
Number of
  revolutions 1000 per min.

	Speed		Tractive force		Hauling power on horizontal track in tons
	m.p.h.	km/h	lbs.	kg	
1st speed	3	4.8	3400	1500	146
2nd ,, ...	7.2	11.5	1540	680	62

The car resistance is taken at 22 lbs. (10 kg) per ton and the adhesion weight at 425 lbs. (190 kg) per ton.

The fuel consumption is 7 pints petrol (4 litres) per brake horse-power per hour. The weight of the locomotor is 8 tons.

LOCOMOTORS OF THE NETHERLANDS RAILWAYS.

The motor shunting locomotives of the Netherlands Railways, made to their own design, occupy a prominent position among traction vehicles of this type. They are built for shunting at small stations and goods yards where there is not enough of this work to keep a locomotive continuously employed the whole day and where a steam locomotive would consequently be uneconomical, as it has to be kept under steam and the driver and fireman have to be kept on duty also in the intervals between shunting operations.

The motor shunting locomotives described below, also called locomotors, were built by the "Berliner Maschinenbau A.G. vormals L. Schwartzkopff" of Berlin.

The considerations underlying the principles of this design are the following:
1) The driver must have a good lookout under all circumstances, also on curved tracks in shunting yards, and it should be possible for the shunter himself to operate the locomotor.

This requirement was met by arranging for the locomotor to be operated from either side and fixing the lowest speed at walking pace. On both sides a wide running board is provided on which the driver can stand and these are placed as low as possible so that the driver can easily jump on while the loco-motor is in motion. Operating handles are naturally also placed on both sides, ena-bling the driver to stand on whichever side affords the better view of the track, and if necessary to go round to the other side when taking a curve in another direction.

2) The coupling of the locomotor to the wagons should be automatic on the buffers coming into contact with each other, so that the driver — who is at the same time shunter — need not leave the locomotor. This is a factor of especial importance when a number of wagons have to be shunted on to factory sidings.

Uncoupling can also be done from the running board.

The locomotor has three speeds, viz. 6, 12 and 30 km/h (4, 8 and 18 m.p.h.). The first two speeds are for shunting and the highest speed is for running light from one station to another.

Where such can conveniently be done, two or three successive stations can be served by one locomotor, in the following way: after doing the necessary shunt-ing at station A and putting the wagons in order ready to be taken along by the next goods train, the locomotor can proceed to station B and do the same there, thence to station C and so on.

When the goods train, moving in the direction A—B—C, arrives at A, all the locomotive of that train has to do is to drop the wagons for that station and take up those set ready by the locomotor, then proceeding to B, and so on to C. In this way the minimum of delay is caused to the goods train.

In the meantime the locomotor returns from C to A and shunts the wagons left by the goods train to the sidings where they have to be loaded or unloaded, doing the same at B and C in succession.

The first locomotor built for the Netherlands Railways had a 30 H.P. petrol engine (see fig. 67), but as this was found to be too low-powered for some stations the next locomotor was built with a 50 H.P. engine, and this gave such satisfaction that it was decided to order several more (see figs. 68 and 69).

Fig. 67. 30 H.P. Petrol Locomotor of the Netherlands Railways.

The main dimensions are:

Engine: number of cylinders 4
         dia. of cylinders . . . . 120 mm $(4^3/_4'')$
         stroke . . . . 180 „ $(7^1/_8'')$
         revs./min 800
         power . . . . 50 E.H.P.
Tractive force: at 6 km/h (4 m.p.h.) . . . . 2000 kg (4400 lbs.)
          „ 12 „ (8 „) . . . . 1000 „ (2200 „)
          „ 30 „ (18 „) . . . . 400 „ (880 „)
Tare . . . . 12½ tons

The engine is a vertical, single-acting, four-cylinder, four-stroke, petrol engine of about 50 H.P. braking power at 800 revs./min. The engine speed can be regulated by means of a throttle valve in the suction line, whilst in order to avoid an excessive number of revolutions a centrifugal governor is installed which also closes a throttle valve in the suction line as soon as 800 revs./min is exceeded.

The engine can be started both by hand and electrically.

Electric high tension ignition is supplied by a Bosch Light Magnet igniter, enabling the engine to be started without a battery by throwing over a crank.

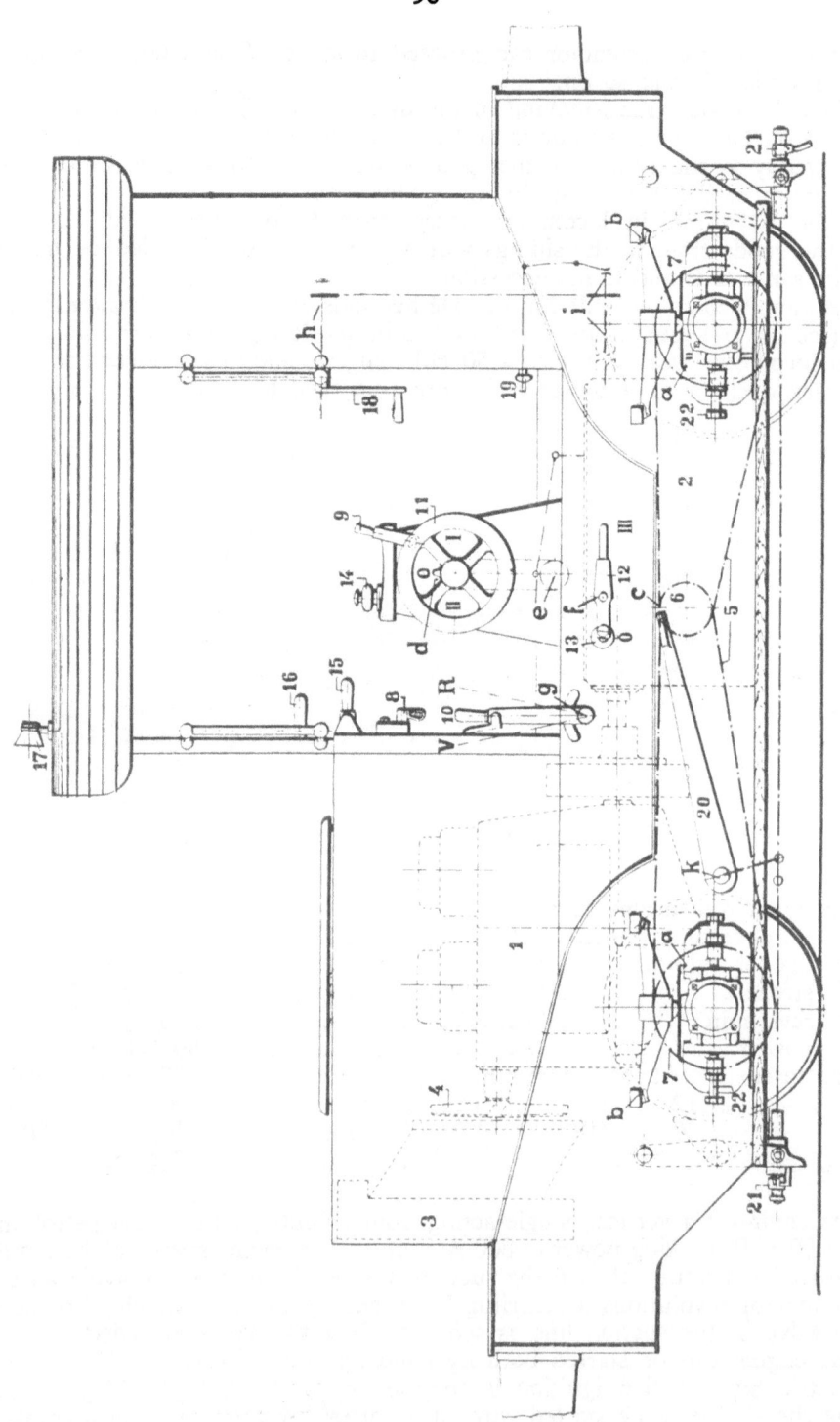

Fig. 68. 50 H.P. Petrol Locomotor of the Netherlands Railways.

Fig. 69. 50 H.P. Petrol Locomotor of the Netherlands Railways.

Lubrication is on the pressure system.

The cooling water for the cylinders is circulated by a centrifugal pump driven by the engine itself, the water being returned to a cooler, in front of which a ventilator is fitted, driven by the engine shaft via gear wheel transmission.

The transmission, see fig. 70, is by gear wheels and chains, acting on both wheel axles. Three variations are possible for either direction of travel, the two lowest

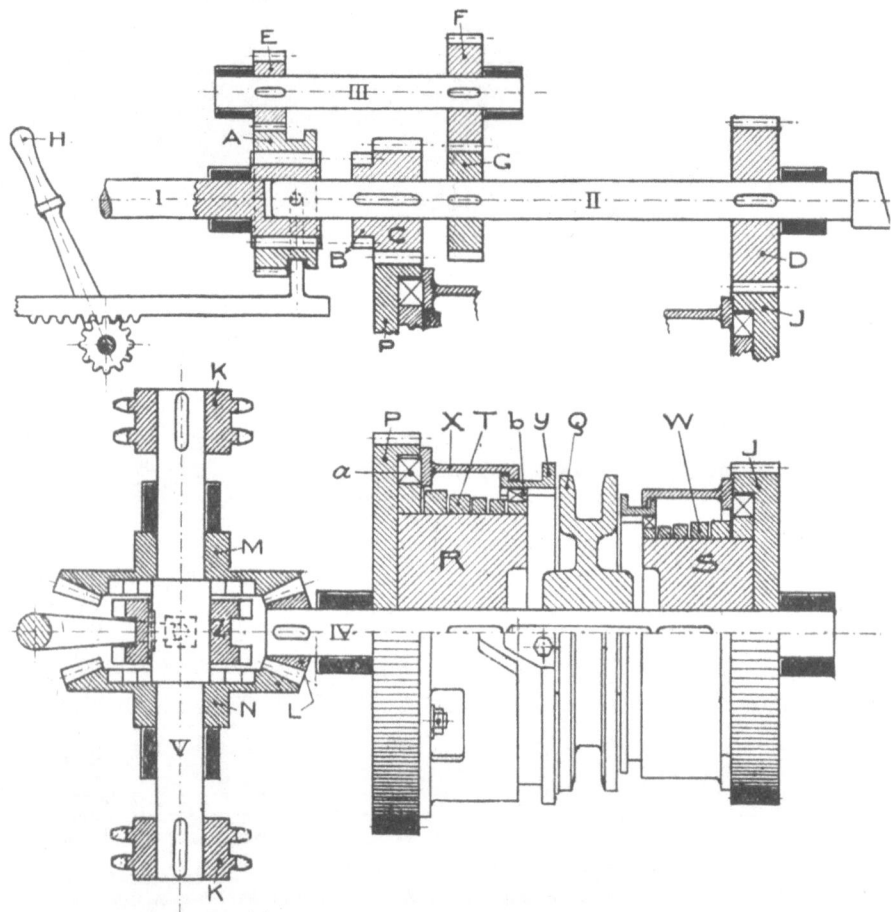

Fig. 70. Diagram of Locomotor of the Netherlands Railways with gear transmission.

of 6 and 12 km/h (4 and 8 m.p.h.) for shunting and the highest of 30 km/h (18 m.p.h.) for running light over long distances. For the latter case the transmission is double-geared, two speed variations being provided, the lower of which (about 12 km/h = 8 m.p.h.), however, corresponds approximately to the highest speed of the single gear action, so that for practical purposes it is of no importance.

In the illustration I is the engine shaft and V the sprocket shaft from which the axles are driven by means of chains. The transmission action is as follows:

When shunting:

Gear wheel A is moved to the right by lever H, causing the internal gear A to engage B, while the external gear A is released from gear E, thus coupling shaft I direct to shaft II.

For the first speed (6 km/h = 4 m.p.h.) the clutch Q is moved to the right, with the following action:

I—A—B—II—D—J—W—S—IV—L—M or N—V.

For the second speed (12 km/h = 8 m.p.h.) clutch Q is moved to the left, the action then being:

I—A—B—II—C—P—T—R—IV—L—M or N—V.

For running light at a higher speed (30 km/h = 18 m.p.h.):

Gear A is placed in the position shown in the diagram, thus releasing it from B and breaking the direct connection between shafts I and II, the transmission between which is then as follows:

I—A—E—III—F—G—II,

the transmission from shaft II to shaft V remaining the same as for the shunting speeds.

For the coupling of gears J and P to the sleeves S and R respectively, which are fixed on shaft IV, so called spring band couplings are provided.

The action of these spring band couplings is as follows: when, for instance, gear P is put into action this rotates, via catch a, the spring band coupling T placed loosely around sleeve R; shaft IV does not then rotate. When, however, clutch Q, which rotates freely on shaft IV, is moved to the left, the ring Y is forced to the left and slides into the ring X, and in consequence of inclined guides in these rings X and Y the latter ring is rotated with regard to ring X, thereby taking along the end b of the spring T and drawing this spring more or less tightly around sleeve R, as a result of which shaft IV is gradually rotated. In this manner shaft IV is coupled to gear P.

On the clutch Q being moved to the right, the gear wheel J is coupled to shaft IV in a similar way.

The direction of travel is regulated by switching the coupling sleeve Z either to the right or to the left, thereby engaging gear wheel M or N with the shaft V.

The whole of the transmission gear is enclosed in a box and runs in an oil bath.

The rotation of shaft V is transmitted to the axles via double chains, which in course of time are elongated by the stretching of the links and wear of the pins, but they are kept taut by extending slightly the distance between the axles, thus increasing somewhat the wheel base. This can be done by removing liners on the outside between journal boxes and box guides and inserting them on the inside.

The frame consists of two heavy frame plates (40 mm = $1^9/_{16}''$ thick so as to give sufficient weight) connected by cross beams, thus forming a rigid whole resting on the journal boxes (with roller bearings) by means of springs.

The wheel axles have separate wheel tyres turned to the standard section.

The brake acts with four shoes on the wheels of both axles, in such a way that the pressure on all shoes is always equal. The brake is operated by a pedal and the maximum brake power is applied by standing on the lever.

Each buffer beam has two standard locomotive buffers, but with lighter springs, and an automatic coupling for the wagons.

This automatic coupling, which was designed in the technical bureaux of the Netherlands Railways, works in the following manner (see fig. 71). Immediately before the locomotive buffers come into contact with the buffers of the wagon or coach to be coupled on, the plate S strikes against the draw-hook of the wagon and the hinge joint of rods A and B (via rod C) is forced back (in the illustration to the left). The three

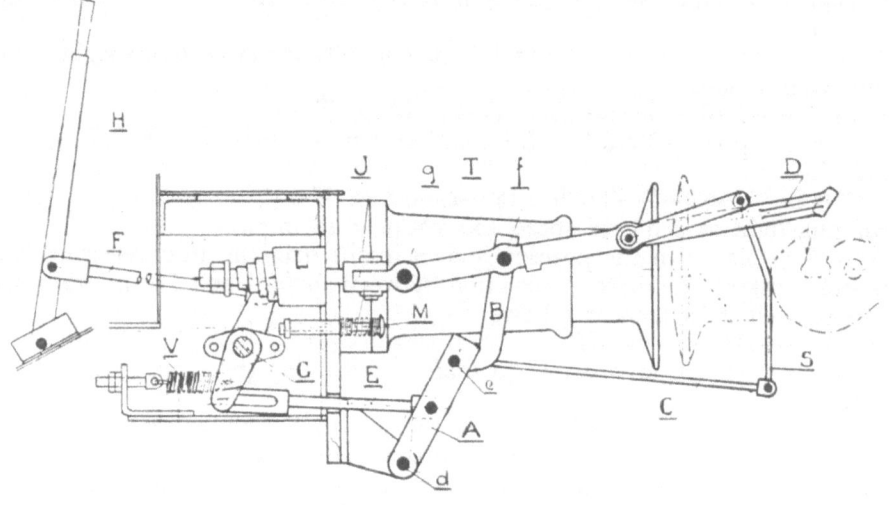

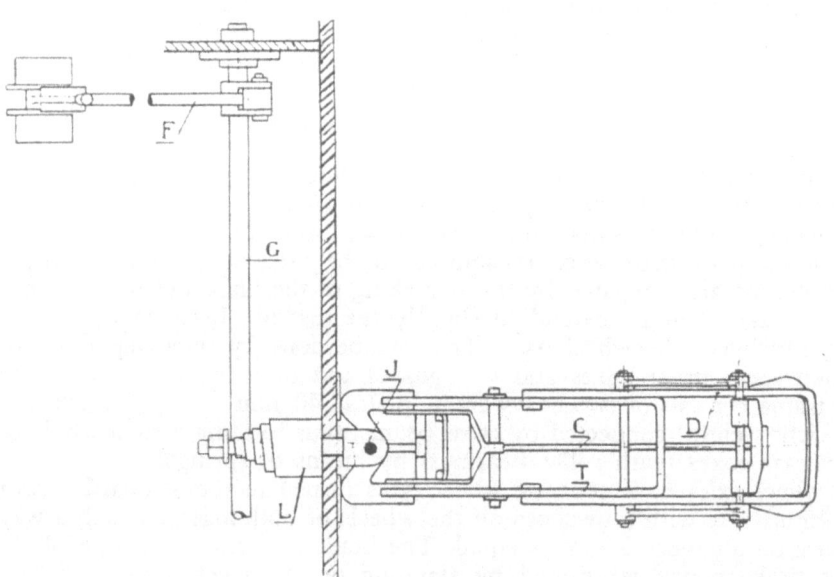

Fig. 71. Automatic wagon coupling, Locomotor of the Netherlands Railways.

hinge joints d, e and f are thrown out of line and the rods A and B allow the draw-link T and coupling link D to fall together, turning on the point g. The coupling link D is then automatically engaged in the draw-hook of the wagon or coach. For uncoupling, at each of the four corners of the locomotor there is an uncoupling lever H with which rod A can be forced back into the position shown in the figure, thereby raising the links T and D and disengaging the latter from the draw-hook.

The following devices have been adopted for giving audible signals:

1) A so-called compression whistle is connected to one of the cylinders and on the respective valve being opened a blast is given at every compression stroke. With this whistle blasts can only be given in quick succession as a general warning signal.

2) A so-called "typhon" by means of which certain signals can be given. One of the engine cylinders is connected to a 100 litre vessel into which the exhaust gas from the cylinder is conducted via a specially water-cooled valve until a pressure of about 5 atm. (75 lbs.) is reached. A safety valve is provided to prevent the pressure rising too high. The "typhon" is connected to the air vessel.

On each side of the locomotor there are the following operating devices, which can be worked from the running board:

a) Handle for reversing.

b) Gas handle for regulating the fuel supply to the engine.

c) Handwheel for varying speed while shunting.

In the normal position this handwheel is placed at O. Speed cannot be changed direct from zero to second speed (12 km/h = 8 miles per hour), the first speed (6 km/h = 4 miles per hour) having to be put in first. On the other hand the wheel can be put back direct to zero from the second as well as from the first speed.

d) High speed handle (on one side only) for putting in the double gear transmission. This interrupts at the same time the engine ignition circuit, and to reconnect this a special button has to be kept pressed in. The idea of this is to ensure that the locomotor will not continue to run at high speed if anything should happen to the driver. When running in third speed (30 km/h = 18 miles per hour) the dead-man's button has to be kept pressed in, as otherwise the engine ignition ceases and the locomotor is brought to a standstill.

e) Foot brake, which can be locked in the braked position.

f) Compression whistle for general warning.

g) "Typhon" for certain signals.

h) Ignition switch operated by a key.

i) Switch for the electric light.

B_0 (0—4—0) DIESEL-ELECTRIC LOCOMOTIVE OF THE NETHERLANDS RAILWAYS.

The Netherlands Railways have rebuilt two electric accumulator locomotives into Diesel electric locomotives, a Diesel engine with generator taking the places of the battery. The expensive accumulator batteries, which had to be renewed after a couple of years' service, are thus dispensed with, whilst moreover the running costs are reduced, the fuel of a Diesel engine costing less than the periodical recharging of the accumulators.

The rebuilding of these locomotives was carried out by the "Berliner Maschinenbau A.G. vormals L. Schwartzkopff" of Berlin. The Diesel engine is an eight-cylinder Ganz engine of 150 H.P. at 1050 revs./min. The generator was supplied by Siemens-Schuckert. It was not necessary to renew the axle motors. The *Gebus* system of transmission is applied.

Fig. 72. Locomotor of the French State Railways.

LOCOMOTOR OF THE FRENCH STATE RAILWAYS.

This locomotor (fig. 72), built by Baudet Donon & Roussel, was destined for shunting service at the St. Lazare station in Paris.

The engine is an explosion engine with 8 cylinders arranged in the form of a V at an angle of 90 degrees; it was built by De Dion Bouton of Puteaux. The "Zenith" carburettor is entirely automatic.

The cooling water is circulated by a centrifiugal pump which forms part of the engine. In front of the cooler a ventilator is provided.

Fig. 73. Interior of Driver's cabin. — Locomotor French State Railways.

The coupling between the engine and the driving mechanism consists of a multiple dry plate coupling on the Hele-Shaw system. The active plates are made of steel and lined with Ferodo, whilst the passive plates are of steel of high tensile strength. These plates are made in a special form ensuring flexibility and avoiding deformation through high temperatures.

The gear-box contains three elements:

a) a carter containing the gear wheels for the various speeds; the gears are always in mesh;

b) the conical gear wheels for reversing;

c) a device for single and double gear, by means of which the number of speeds can be doubled.

The axles of the locomotor are driven with the aid of double roller chains (pitch 50.8 mm).

There is a complete Westinghouse brake and a hand brake, as also a sand box for either direction of travel.

The engine is started with compressed air supplied by the same compressor as provides the air for the Westinghouse brake. Should the air vessels be empty the engine can also be started by hand.

A double set of handles is fitted up in the driver's cabin (one set at each side) so that the driver can always stand where he has the best lookout. The cabin is entirely closed in and the various levers etc. are lighted without hindering the driver's view. Fig. 73 shows the interior of the cabin.

The main dimensions of this locomotor are:

Engine
- bore . . . . 125 mm ($4^5/_{16}$")
- stroke . . . . 150 „ ($5^7/_8$")
- revs./min 1000
- power . . . . 100 H.P.

Capacity of petrol tank . . . . 250 litres (55 glns.)
„ „ cooling water tank . . . . 1000 „ (220 „)
Number of speeds 12
Tractive force — at lowest speed (2 km/h) ($1^1/_4$ m.p.h.) . 10000 kg
— at highest speed (60 km/h) (37 m.p.h.) . 400 kg
Weight in working order . . . . 30 tons

Fuel consumption is about 10 litres (2.2 glns.) per shunting hour, i.e. with frequently interrupted working during an hour.

LOCOMOTOR OF THE PARIS—LYONS—MEDITERRANEAN RAILWAY.

Fig. 74. Locomotor of the Paris—Lyons—Mediterranean Railway.

This locomotor is shown in fig. 74 and was built by "Les Automobiles M. Berliet" at Lyons. The frame, on which the engine with accessories and the driver's cabin are mounted, is very similar to that of an ordinary coach. The engine is an explosion engine, the power of which is transmitted to the two driving axles by gear wheels and chains. Lubrication is effected under pressure. The engine has a speed governor and a radiator with ventilator is provided for cooling the cylinders; the cooling is circulated by a centrifugal pump.

The engine can be started electrically or by hand. The coupling between engine and driving gear is by means of a friction clutch consisting of a series of bronze and steel discs. The driving chains are kept taut by adjusting the journal boxes between the box guides.

In the cabin there are handles for the following purposes, placed on both sides:

a) for reversing,
b) for coupling,
c) for changing speed,
d) for acceleration.

An accumulator battery, automatically charged by a dynamo, provides for lighting and the starting of the engine.

A capstan is mounted on the frame for shunting wagons on an adjacent track with the aid of a cable.

There is a double set of brakes, a rapid-acting brake worked with a handle and acting on the sprocket-shaft with brake drums, and an ordinary screw-brake with shoes bearing on the wheels.

The main dimensions of this locomotive are:

Engine
- power . . . . 40 H.P.
- cylinders 4
- bore . . . . 110 mm ($4^5/_{16}''$)
- stroke . . . . 140 „ ($5\frac{1}{2}''$)
- revs./min. 1500

Speeds
- 1st gear . . . . 4.45 km/h ($2^3/_4$ m.p.h.)
- 2nd „ . . . . 9.15 „ ($5^3/_4$ „ „)
- 3rd „ . . . . 20.2 „ ($12\frac{1}{2}$ „ „)
- 4th „ . . . . 33.3 „ ($20\frac{1}{2}$ „ „)

Max. tractive force of capstan . . . . 1500 kg (3300 lbs.)
Length of frame . . . . 5.50 m ($18'$ $0\frac{1}{2}''$)
Length over buffers . . . . 6.63 „ ($21'$ $9''$)
Width . . . . 2.98 „ ($9'$ $9^1/_4''$)
Height from top of rails . . . . 3.37 „ ($11'$ $0^3/_4''$)
Wheel base . . . . 2.70 „ ($8'10^1/_4''$)
Wheel diameter . . . . 0.93 „ ($3'3^5/_8''$)
Weight without ballast . . . . 12 tons
„ with „ . . . . 17 „

The figures for the tractive force and fuel (petrol) consumption per km are:

	Speed		Tractive force		Petrol consumption per km	
	km/h	m.p.h.	kg	lbs.	litres	pints
Starting	—	—	2400	5300	—	—
1st	5	3	1000	2200	2	3.5
2nd	10	6	500	1100	1	1.76
3rd	20	12	250	550	0.5	0.88
4th	30	18	170	375	0.3	0.53

SCHNEIDER & CO LOCOMOTORS.

Messrs. Schneider & Co. of Paris build locomotors of 15 and 22 tons for normal gauge and some of these are in service on the Chemin de fer du Nord (France). The 15 ton locomotor is intended principally for shunting on factory sidings, while the 22 ton type is for main line service (the smaller type is shown in fig. 75). Transmission is by means of gear wheels and chains working on the two driving axles.

Fig. 75. Locomotor of Schneider & Co.

The 22 ton locomotor (fig. 76) has air and hand brakes, whereas the 15 ton one has only a hand brake. The main dimensions are:

		15 tons	22 tons
Tare		15 tons	22 tons
Engine	power	55	80 E.H.P.
	cylinders	4	4
	bore	135 mm (5⁵⁄₁₆″)	135 mm (5⁵⁄₁₆″)
	stroke	170 mm (6¹¹⁄₁₆″)	170 mm (6¹¹⁄₁₆″)
	revs./min	1000	1400
Length over buffers		5.20 m (17′0³⁄₄″)	6.60 m (21′7¹⁄₈″)
Wheel base		1.80 m (5′10⁷⁄₈″)	2.60 m (8′6³⁄₈″)

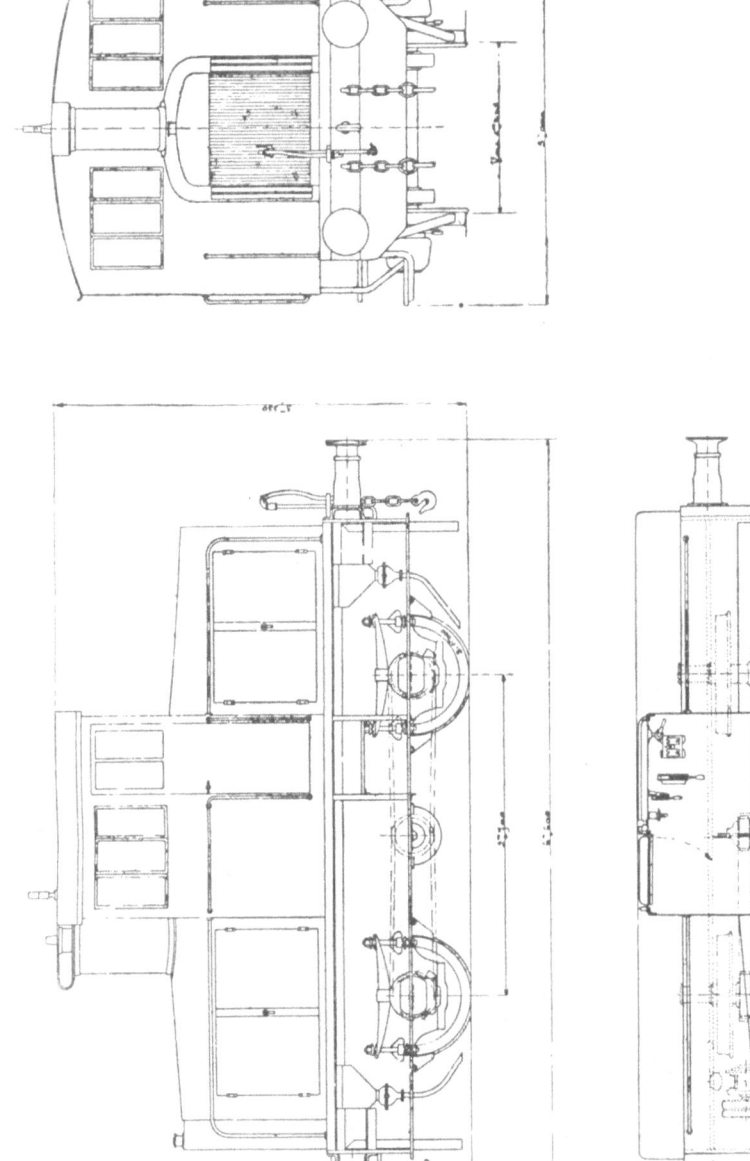

Fig. 76. Locomotor of Schneider & Co.

15 ton locomotor:

	Speed		Tractive force at draw-hook		Hauling capacity in tons
	km/h	m.p.h.	kg	lbs.	
1st speed	4.3	2	3000	6600	240
2nd ,, ...	7.5	4	1800	3950	145
3rd ,, ...	11.2	7	1150	2500	90
4th ,, ...	16.2	10	750	1650	60

The wagon resistance is calculated on a level track at 10 kg (22 lbs.) per ton.

22 ton locomotor:

	Speeds			Tractive force	
	Number	km/h	m.p.h.	kg	lbs.
single gear	4	4—13	2½—8	4800—1600	10600—3500
double gear	4	20—60	12½—37	1100—330	2400—750

Petrol consumption:

at full load. . . . . 0.4 litres (0.7 pints) per H.P.h
,, $^3/_4$,, ... 0.44 ,, (0.78 ,,) ,, ,,
,, $^1/_2$,, ... 0.52 ,, (0.92 ,,) ,, ,,

DIESEL-ELECTRIC LOCOMOTIVES OF THE "DIESEL ELEKTRISKA VAGN- AKTIEBOLAGET VÄSTERÅS", SWEDEN.

The above-named works build Diesel-electric locomotives of the following main dimensions:

Engine power in H.P. 120 150 200 250 300 500
Tare in tons 25 28 37 45 48 78
Max. speed in km/h 60 60 60 60 60 80
,, ,, ,, m.p.h. 37 37 37 37 37 49
Illustration fig. No. 77 78 79 80 81 82

These locomotives are in use on the following railways:

Railway	Number	Rating H.P.	Year taken into service	Gauge			Vide fig.
				Metres	ft.	in.	
Sweden. Halmstad—Nässjö Railway	2	150	1921	1.435	4	8½	83
	3	200	1925	1.435	4	8½	
Skåne—Smålands Railway	1	150	1921	1.435	4	8½	
Norrköping—Söderköping Vikbo-landets Railway.	1	120	1917	0.891	2	2	
	1	120	1922	0.891	2	2	
	1	150	1925	0.891	2	2	
Hälsingsborg—Hässleholms Rail-way.	1	120	1919	1.435	4	8½	
	1	250	1921	1.435	4	8½	

Railway	Number	Rating H.P.	Year taken into service	Gauge			Vide fig.
				Metres	ft.	in.	
Upsala—Enköpings Railway	1	300	1922	1.435	4	8½	84
Mellersta Östergötlands Railway	1	75	1923	0.891	2	2	
	1	120	1923	0.891	2	2	
Denmark. The Lolland—Falstersko Railway	1	120	1923	1.435	4	8½	
Helsingör—Hornbaek—Gilleleje Railway.	1	120	1923	1.435	4	8½	
Amager Railway	1	150	1924	1.435	4	8½	
State Railways	1	400	1929	1.435	4	8½	
Tunis. Cie. Fermière des Ch. d. f. tunisiens	1	120	1923	1.000	3	3	89

The driving power is supplied by a four-cycle Diesel engine with 4 to 12 cylinders, coupled direct to a generator supplying the electric current which is transmitted to the electric-motors, which in turn transmit the power to the axles via

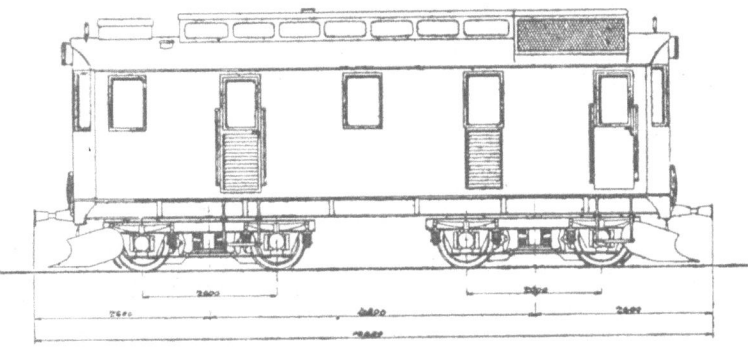
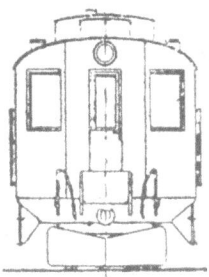
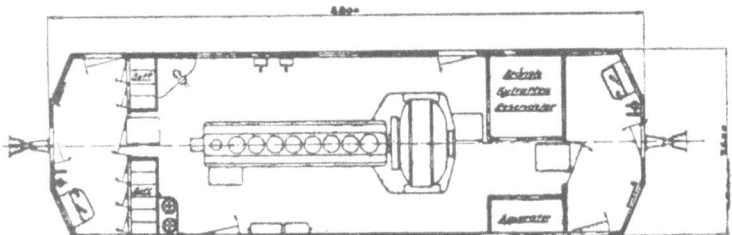

Fig. 77. 120 H.P. "Deva" narrow gauge Diesel-electric locomotive.
Apparater = Apparatuses. — Batt = Battery. — Bransle = Fuel. — Kylvatten = Cooling water. —
Reservdeler = Spare parts. — Spv = Gauge of track.

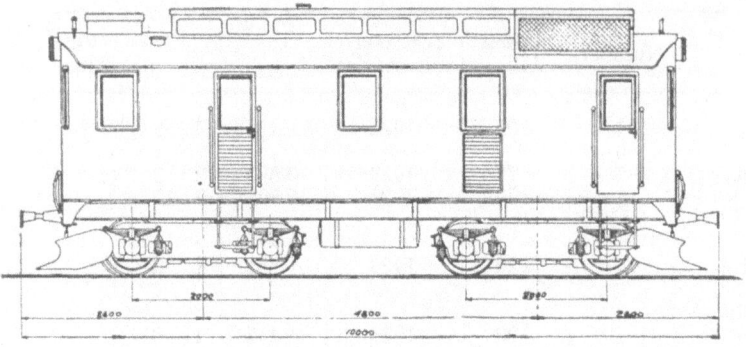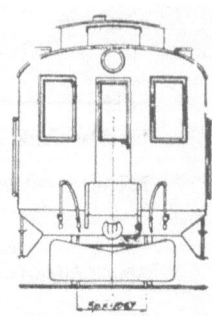

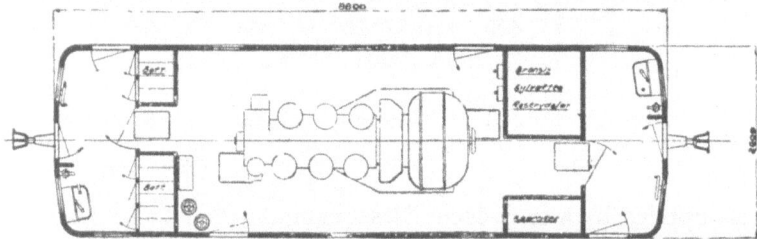

Fig. 78. 150 H.P. "Deva" narrow gauge Diesel-electric locomotive.
Apparater = Apparatuses. — Batt = Battery. — Bransle = Fuel. — Kylvattan = Cooling water.
Reservdeler = Spare parts. — Spv = Gauge of track.

mechanical gearing. The Diesel engine and generator are mounted on one frame and connected up by a flexible coupling.

The fuel is injected into the Diesel engine by a pump, thus without compressed air. It is distributed in the cylinders by means of special distributors, in such a way that if necessary one or more cylinders can be put out of action if anything should happen to one of them, at the same time allowing the engine to be run on the other cylinders.

All parts are lubricated under pressure. Excess oil is collected in the bottom of the crank case, from which it flows via a filter to a tank placed under the floor of the engine room; from this tank the oil is pumped to the main lub.oil tanks.

The parts which are exposed to high temperatures, such as the cylinders, cylinder covers, etc., are cooled with water circulated by a pump driven direct from the engine. The cooling water is recooled by a system of radiators and tubes mounted on the roof, assisted by ventilators.

The voltage of the current supplied by the generator may vary between very wide limits, up to 750 volts; the generator shaft is mounted on ball or roller bearings of the S.K.F. make.

The generator is of the eight-pole type with commutator poles. It is shunt wound and provided with a separate series of coils which are in circuits when the generator is used as motor driven from an accumulator battery.

The generator employed for the propulsion of the locomotive is thus likewise used for starting the Diesel engine and also for charging the accumulator battery.

The electric motors are entirely enclosed; in the shell is a cover through which the brushes and collectors can be inspected. The motors are suspended on the axle

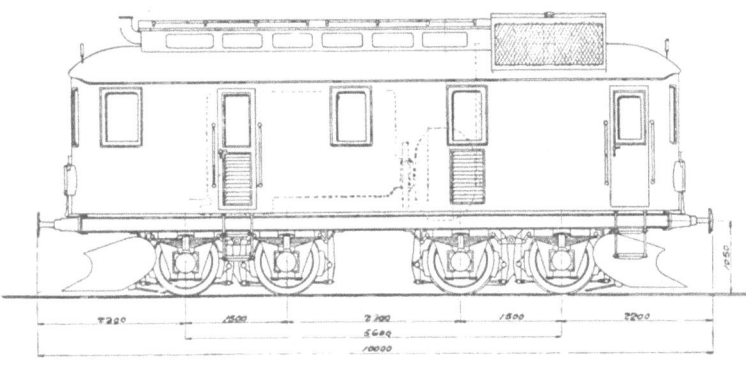
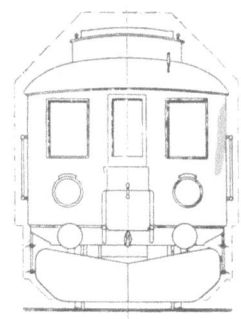

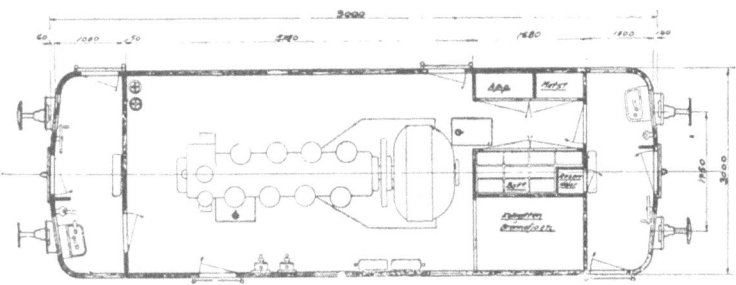

Fig. 79. 200 H.P. "Deva" Diesel-electric locomotive.

App = Apparatuses. — Batt = Battery. — Reservdeler = Spare parts. — Brännolje etc. = Fuel, etc.
— Motst = Resistance. — Kylvatten = Cooling water.

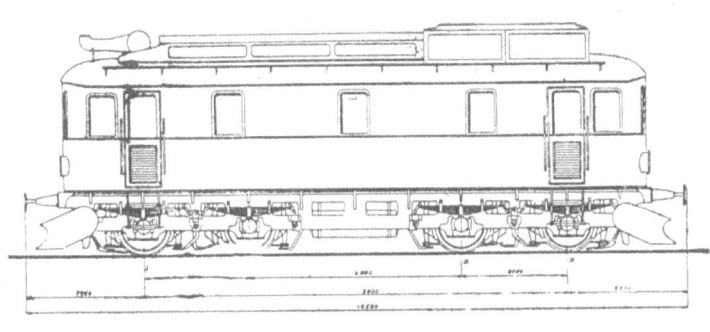
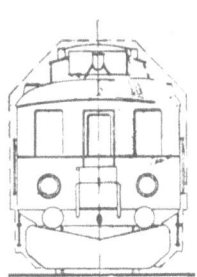

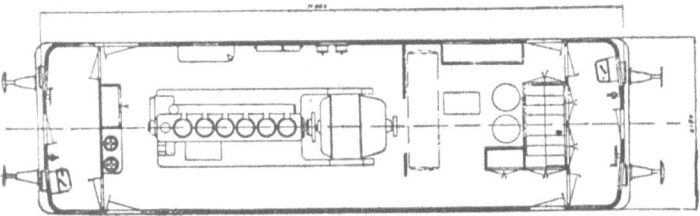

Fig. 80. 250 H.P. "Deva" Diesel-electric locomotive.

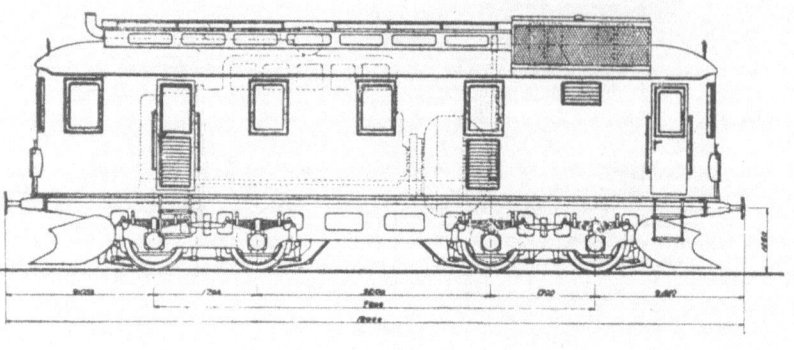
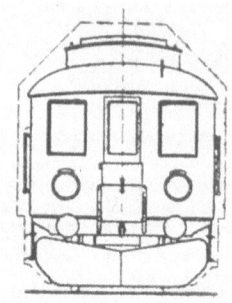

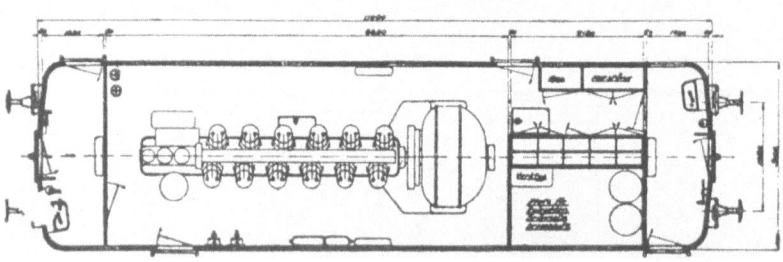

Fig. 81. 300 H.P. "Deva" Diesel-electric locomotive.

App = Apparatuses. — Brännolje = Fuel. — Bremsluft = Air for brake. — Kylvatten = Cooling water. — Motständ = Resistance. — Plats for = Room for. — Verktyg = Tools.

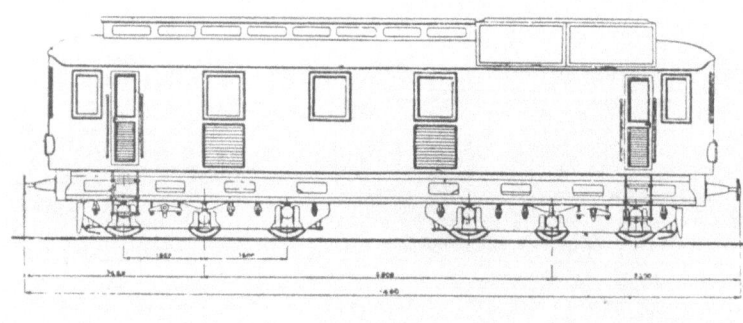

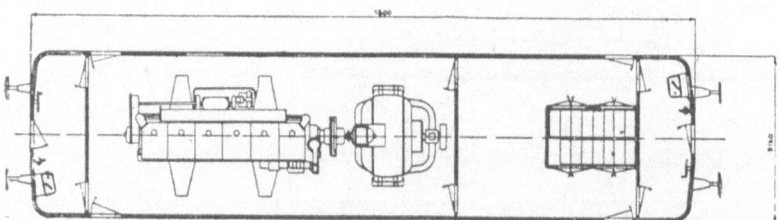

Fig. 82. 500 H.P. "Deva" Diesel-electric locomotive.

on two bearings and on the other side are fixed to the locomotive frame. The springs are mounted so as to swing free, thus enabling the motors to follow the movement of the shaft in respect to the frame. The gearing is made of chrome nickel steel, is entirely enclosed and rotates in oil.

The traction motors are cooled by ventilators mounted on the rotor shaft. The cooling air is

Fig. 83. 150 HP Diesel-electric locomotive of the Halmstad—Nassjo Railway.

drawn in at the top of the box and conducted to the air vessels through flexible tubes.

By means of the switches placed in the drivers' cabin a motor can be disconnected in the event of it breaking down, in which case the locomotive can still be run on the other motors.

The two operating stands are fitted with all apparatuses and instruments required for running and controlling the engines, with levers for the vacuum or air brake, whistle, sand-boxes, wattmeters recording the power produced, manometers for air pressure and oil pressure in the bearings of the Diesel engine, and a speed indicator.

The controls consist of two cylinders or drums, one for the forward and reverse direction and the other for regulating the speed of the locomotive. The lever of the reversing drum can only be used when the speed regulator handle is in the neutral or zero position. When the reversing lever is placed in the neutral position the speed regulator handle is locked, thus forming a safeguard against wrong manipulations.

The drum for speed control sets the engine in motion, connects up the motors and regulates the speed of the locomotive; the handle is fitted with a dead man's grip, and on this being released, e.g. if the driver meets with an accident, the engine stops immediately, the locomotive slows down and finally comes to a standstill, the brakes being applied automatically.

The continuous current required for starting the Diesel

Fig. 84. 300 HP Diesel-electric locomotive of the Upsala—Enköpings Railway.

8

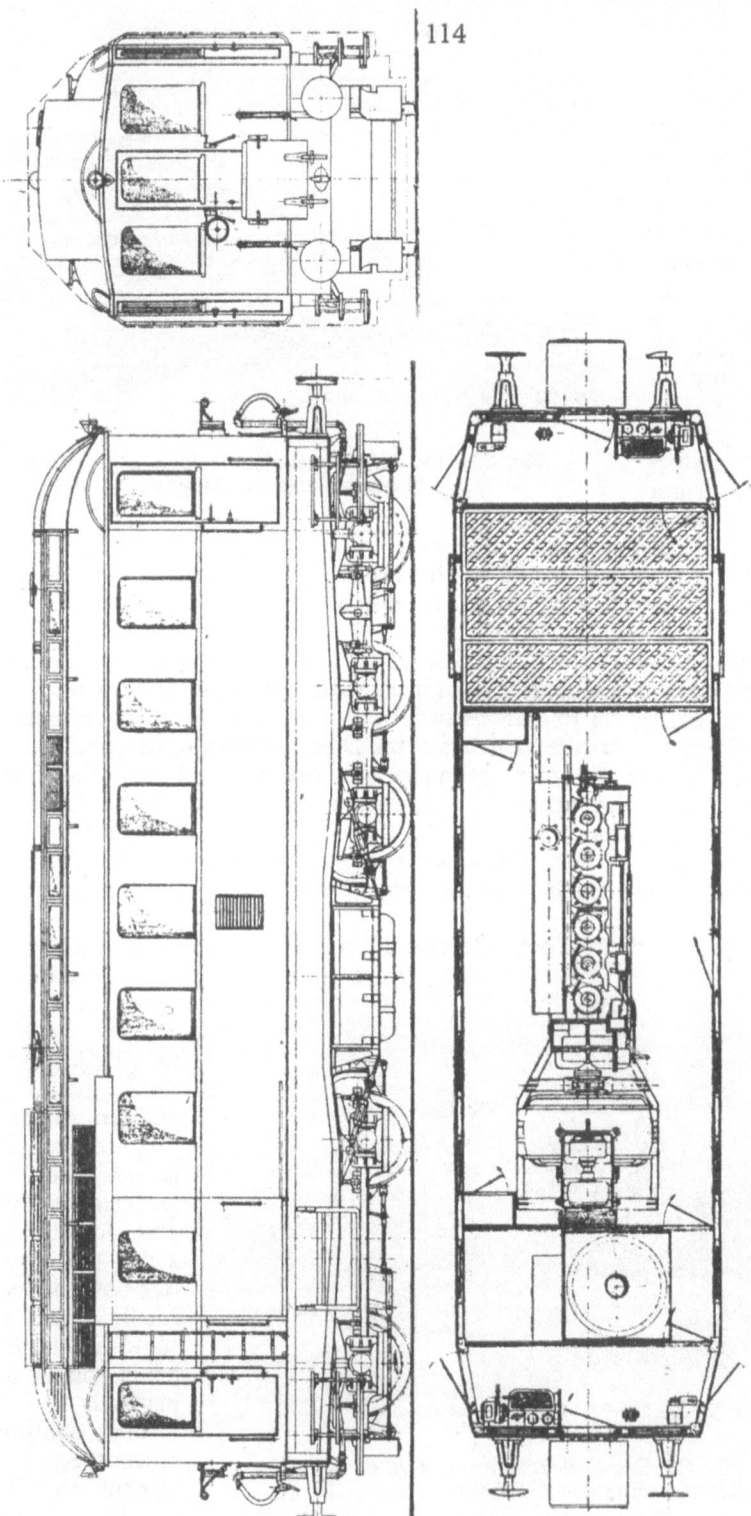

Fig. 85. 450 H.P. Burmeister & Wain Diesel-electric locomotive.

engine, for the lighting and for the operation of various apparatuses is supplied from an accumulator installed in the locomotive. This is a *Jungner* alkali battery and while the locomotive is running is automatically charged; in the engine room is a special switch for charging the battery while the locomotive is stationary.

The cooling water for the motor is cooled by a radiator system with ventilators placed on the roof of the locomotive; the ventilators are driven direct from the driving shaft and as this rotates continuously during the run the functioning of the ventilators is consequently not dependent on the speed of the train.

There are double brakes, a hand brake and a vacuum or compressed air brake; the vacuum or air pump is driven by a small electric motor.

If the accumulator is of adequate dimensions the train can also be lighted from the locomotive.

BURMEISTER & WAIN'S DIESEL-ELECTRIC LOCOMOTIVE.

The "Aktieselskabet Burmeister & Wain's Maskin go Skibsbyggere" of Copenhagen have built a Dieselelectric locomotive as shown in fig. 85. The Diesel engine, built by the same firm, is a single-acting six-cylinder, compressorless engine rated at 450 H.P. at about 550 revs./min. It is coupled direct to a direct current generator supplying current of 650 volts tension. There is no flexible coupling and no flywheels, the engine being accurately balanced and able to answer to the elastic deformations of the locomotive. An exciter dynamo and a dynamo for charging the battery are driven from the engine shaft by means of a belt.

The Diesel engine is started by running the generator as a motor on current from a battery of accumulators consisting of 33 cells, with a capacity of 500 Ah for a time of discharge of five hours.

There are five compartments in the locomotive:
in the middle the engine room, a boiler room and a luggage compartment, and at each end a driver's cabin with a door and a gangway leading to the next coach.

Fig. 86. Diesel-electric locomotive of the Danish State Railways.

The main dimensions are:
Length over buffers . . . . 12.17 m (39'11$^1/_8$")
Length over body . . . . 10.90 m (35'8$^7/_8$")

Greatest width	2.90 m	(9' 6¼")
Dia. of driving wheels	1054 mm	(3' 5½")
Dia. of trailers	966 „	(3' 2")
Gauge	1435 „	(4' 8½")

Fig. 87. Train with Diesel-electric locomotive of the Danish State Railways.

For cooling the circulation water and the lubricating oil a cooler with two ventilators is mounted on top of the luggage compartment. The driving motors, which are of the enclosed tramcar type, drive the front and rear axles.

Vacuum and air brakes are provided.

During the trial runs a max. speed of 60 m.p.h. (96 km/h) was reached with a train weight of 20 tons (excl. passengers and luggage).

Illustrations of this locomotive are given in figs. 86 and 87.

FRICHS' DIESEL LOCOMOTIVES.

The firm of Frichs at Aarhus (Denmark) finished their first Diesel locomotive in the beginning of 1926; it was built for a small private railroad Hads-Ning-Herreders for the line Aarhus-Odder-Hou.

This type of locomotive has the following dimensions:

Weight . . . 38 tons
Length over buffers
11550 mm (37' 10³/₄")
Wheel base
6620 mm (21' 8⁵/₈")

Diameter of driving wheels . . . . 996 mm (3' 2")
Fuel storage . . . . 350 kg (775 lbs.)
Max. tractive effort . . . . 4500 kg (10000 lbs.)

This locomotive is of the A 1 + 1 A (0—2—2 + 2—2—0) type; it has control cabins at both ends and a luggage compartment. The Diesel engine has 6 cylinders, runs with 500 revs. p. min and has a max. rating of 250 H.P.

Since the first day it was taken into service this locomotive has given satisfaction; its daily duty was to make 10 trips between Aarhus and Hou, totalling some 375 km and up to the time of writing it had travelled more than 500,000 km almost twice the distance covered by the steam locomotive which it replaced; it is attended to by one man and gives a considerable saving in fuel.

In the first year, when coal prices were pretty high, the saving in fuel amounted to more than 30,000 Danish Kroner, and even with the present low prices the saving is still more than 20,000 D.K. per year. The running expenses were also lower than those of a steam locomotive and now, after a period of 5 years, a saving of 25 % has been realised.

Other private railways soon purchased several locomotives of the same type, in sizes up to 325 H.P. The Danish State Railways have also been interested for years in Diesel operation and after the

Fig. 88. Frichs' 450 H.P. Diesel-electric locomotive Danish State Railways.

satisfactory results obtained by the private railways they decided to buy six small Diesel locomotives of the same type as described above, but with two bogies instead of rigid axles.

In 1928 the State Railways ordered two more locomotives of 450 H.P.; a brief description of this typical Frichs' locomotive, which has been in service 2 years already, will be given below.

One of these 450 H.P. Diesel locomotives, type 1 B$_0$ 2 (2—4—4) is shown in fig. 88. They are now in service on the Tønder-Sønderborg line, each 12 hours per day, covering a distance of 350 km with a speed of 45 km/h, this being the speed limit for that district. They are, however, constructed for a speed of 80 km/h, and on their trial runs on the Jutland main line they reached a max. speed of 107 km/h, the engine still running very smoothly. It is intended to use exclusively Diesel traction in that district and for this reason the engines were transferred from the Seeland district (Copenhagen-Korsør and Copenhagen-Gedser) to above-named line.

The main dimensions for this type are:

Weight . . . . 55 tons
Length over buffers . . . . 11,490 mm (37'8³/₈")
Total wheel base . . . . 7,700 „ (25'3¹/₈")
Diameter of driving wheels . . . . 1,400 „ (4' 7")
    „   „ trailers . . . . 934 „ (3' ³/₄")
Fuel storage . . . . 600 kg (1325 lbs.)
Max. tractive effort . . . . 5,500 kg (12000 lbs.)

These locomotives were built with heavy frames, just like steam locomotives, and the entire superstructure has been executed in steel and in light metal. In the engine room is a vertical steam boiler for heating the train; this is automatically fired either with fuel oil or by the exhaust gases of the Diesel engine. The locomotive is operated by one man; the master valve is fitted with a dead-mans' grip, which, when released, causes the brake to be applied immediately and the engine stopped. Fig. 89 shows the operating stand.

The shunt rheostat is controlled by the main valve lever, whilst the small lever may be turned into three positions serving to change the driving motors into series or into parallel motors. The small lever, which is not clearly visible in the illustration, is for regulating the number of revolutions. There are various instruments on the operating stands: brake-manometer, speed gauge, red and green control lamps for the master valve, cooling water thermometer, apparatus for controlling the pressure of the lubricating oil and the cooling water, amperemeter and voltmeter.

Fig. 89. Frichs' Diesel-electric locomotive Danish State Railways, Driver's Cabin.

The locomotive frame consists of heavy longitudinal girders reinforced by heavy plates and steel cross beams, on which the engine rests. The driving axles rotate on roller bearings and the other axles on journals. The springs of the driving wheels are connected by compensating levers. The superstructure, made of structural and sheet-steel, consists of 3 parts: control cabin, engine room and combined control cabin and luggage room. The roof of the engine room may be taken apart in two parts; the sidewalls are also collapsible. The engine room contains, besides the engine, a vertical oil-fired boiler for heating the locomotive and the train in winter.

The boiler is automatically fired and does not require special supervision.

During the trial run in May 1929 on the Jutland main line, with a train consisting of 3 bogie carriages (total load 120 tons), the locomotive travelled on schedule time; when the speed was increased to 107 km/h by way of experiment the locomotive ran just as smoothly as at the required speed of 80 km/h.

The fuel consumption was about 0.74 kg/km with an average load of 100 tons.

Fig. 90 is a diagram showing gradients etc. of the trial run.

The Danish State Railways then ordered two more Frichs' locomotives of 900 —1000 H.P. for express service in Jutland from Padborg to Frederikshavn. On this

450 km line there are three termini, so that with steam operation several locomotives were required, whereas the new Diesel locomotive with its driver's cabins at both ends does not have to be turned and can carry sufficient fuel for the whole distance, enabling it to cover the whole journey with only a few minutes' delay at the termini. Each locomotive has 2 Diesel engines of 450/500 H.P. each at 600 revs. p. min.

The cooling plant for cooling the water consists partly of natural cooling on the roof and partly of a reserve installation with ventilators.

These locomotives are also provided with an oil-fired boiler for heating the train.

As the Danish laws require that all passenger coaches must have a vacuum brake and all transito freight cars compressed air brakes, the locomotives are fitted with both systems.

These last locomotives are of the 2—D₀—2 type (4—8—4) having 4 driving axles driven by electric motors with gear transmission; they are built for express service with a load of 250 tons at a speed of 100 km/h.

These locomotives, like those of 450 H.P. mentioned above, have a steel frame running over the entire length,

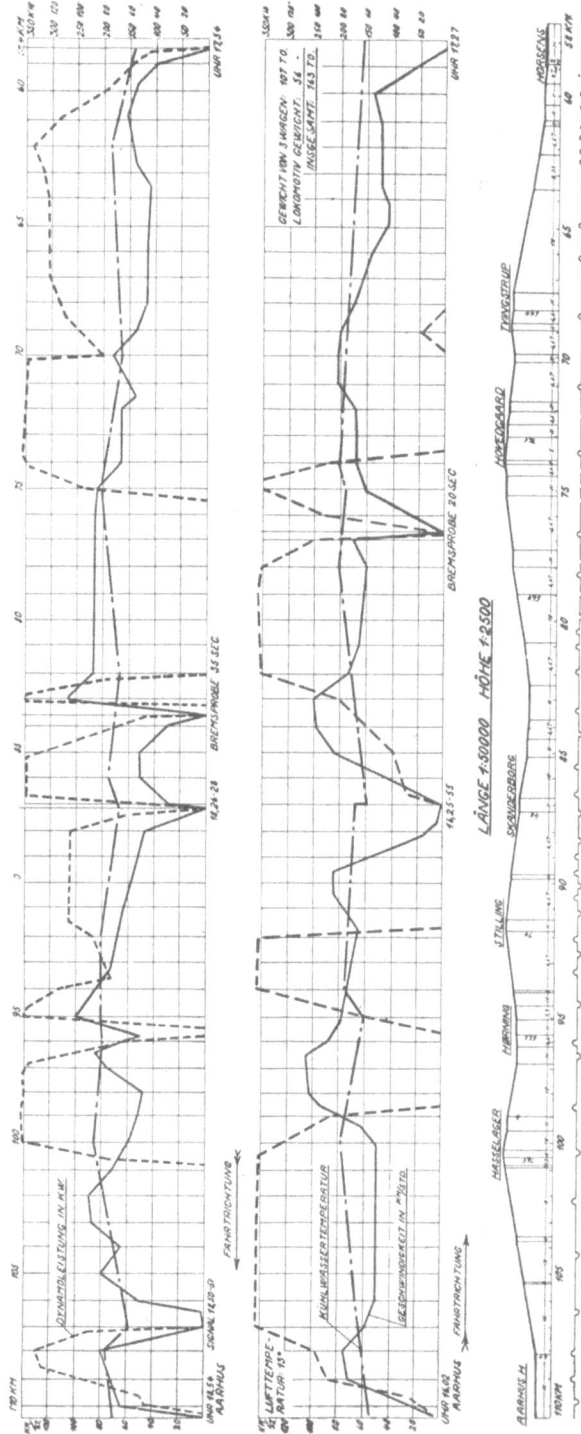

Fig. 90. Frichs' 450 H.P. Diesel-electric locomotive Danish State Railways. Diagram of trial run.
Bremsprobe = Brake test. — Dynamoleistung in KW = Dynamo output in kilowatt. — Fahrtrichtung = Direction of running. — Geschwindigkeit in KM/STD = Speed in kilometres per hour. — Höhe = Height. — Insgesamt = Total. — Kühlwassertemperatur = Temperature of cooling water. — Länge = Length. — Lokomotivgewicht = Weight of locomotive. — Lufttemperatur = Temperature of atmosphere. — Uhr = Time.

with the Diesel engines with directly coupled generators resting on the cross beams.
Their main dimensions are:

Weight	. . . .	98 tons
Length over buffers	. . . .	17,020 mm (55' 10")
Total wheel base	. . . .	12,190 „ (40' 0")
Diameter of drivers	. . . .	1,404 „ (14'7$^{1}/_{4}$")
„ „ trailers	. . . .	934 „ (3' $^{3}/_{4}$")
Fuel storage	. . . .	2,500 kg (5500 lbs.)
Max. tractive force	. . . .	11,000 „ (24200 lbs.)

When the locomotives are in continuous service the rule is that two hours per week is to be allowed for cleaning, inspection and grinding of valves etc. In accordance with the Danish Railway regulations, a complete overhaul takes place after a distance of 50,000 km has been travelled, but it was proved that with these locomotives this distance can easily be increased and the State Railways have now fixed the figure at 70,000 km; some railways have taken 90,000 km and believe that this may even be increased to 100,000 km.

The high speed Frichs' Railway Diesel engines run with a moderate number of revolutions, viz. from 600 to 1000 p. min, according to their sizes. No special efforts have been made to reduce their weights per H.P.

FRICHS' LOCOMOTIVES FOR THE SIAMESE STATE RAILWAYS.

The Siamese State Railways have one 1400/1550 H.P. locomotive and six 900/1000 H.P. Diesel electric locomotives for a gauge of one metre.

The first named is a goods locomotive for the Bangkok-Chingmai division and covers on the up and down trips a total of 1500 km (900 miles) without any stops for taking in fuel, water, etc. It is of the 2 D_0 + D_0 2 (4—8—0 + 0—8—4) type with bogies at both ends and 16 driving wheels driven by electric motors with gear wheel transmission.

Its main dimensions are:

Weight	. . . .	128 tons
Length over buffers	. . . .	21,584 m (70' 9$^{3}/_{4}$")
Total wheel base	. . . .	17,780 „ (58' 4")
Diameter of drivers	. . . .	1,050 „ (3' 9$^{1}/_{4}$")
„ „ trailers	. . . .	762 mm (2' 6")
Fuel storage	. . . .	5 tons (11000 lbs.)
Max. tractive effort	. . . .	22 tons (48500 lbs.)

The locomotive has two 775 H.P. Diesel engines coupled direct to generators and rotating normally with 600 revs. p. min. Both the frame and the superstructure are built entirely of steel and in view of its great length the locomotive is built in two halves articulated in the centre.

The six 900/1000 H.P. locomotives are built for a speed of 60 km/h (37 m.p.h.) and run on the line Bangkok—Penang on the Malakka Peninsula. The distance of 1000 km (600 miles) is covered without change of engine; sufficient fuel oil for the return trip is carried. For the quickest and most modern trains in Siam, carrying mail and passengers from European steamers from Penang to Bangkok and vice versa, the trip takes 20 hours.

The train consists of 13 carriages, mostly first-class saloon, sleeping and dining cars.

With the former steam traction about 6 hours were required for taking in fire-wood and water; this time is now saved.

These locomotives are of the 2 D_0 2 (4—8—4) type with two bogies at the ends and eight driving wheels driven by electric motors with gear wheel transmission. Their main dimensions are:

Weight	88 tons (194,000 lbs.)
Length over buffers	15,390 m (50′ 60″)
Total wheel base	10,900 m (35′9$^1/_8$″)
Diameter of drivers	1,050 m (3′5$^3/_8$″)
,, ,, trailers	762 mm (2′ 6″)
Fuel storage	5 tons (11,000 lbs.)
Max. tractive effort	13 tons (28,600 lbs.)

All these locomotives have coolers, mounted sideways, and mechanically driven ventilation. Driver's cabins are at each end and are designed in every respect for service in the tropics.

DIESEL-ELECTRIC LOCOMOTIVE FOR JAPAN.

This locomotive (see fig. 91) was built by the "Maschinenfabrik Esslingen". It has a MAN Diesel engine coupled direct to a DC shunt dynamo of 750 volts, which is excited by an auxiliary generator. The electric equipment was made by Brown Boveri & Cie of Baden.

The Diesel engine has six single-acting cylinders and is compressorless. It is started with compressed air delivered by a compressor driven by the main engine. There are two electric motors, both driving a loose axle coupled to the driving wheels by connecting rods.

The main dimensions of the locomotive are:

Engine power	600 E.H.P.	
Dia. of driving wheels	1250 mm	(4′ 1$^1/_4$″)
,, ,, bogie wheels	940 ,,	(3′ 1″)
Rigid wheel base	3000 ,,	(9′10$^1/_8$″)
Total wheel base	7550 ,,	(24′ 9$^1/_4$″)
Length overall	10650 ,,	(34′ 11$^1/_4$″)
Gauge	1067 ,,	(3′ 6″)
Weight in working order	58.3 tons	
Max. speed	60 km/h (37 m.p.h.)	

Tractive force at driving wheel periphery at a speed of
18 km/h (11 m.p.h.). . . . . 7500 kg 16500 lbs.)

The locomotive can set in motion a train of 500 tons (excl. the locomotive) on a gradient of 3 : 1000 and reach a speed of 25 km/h (15 m.p.h.). On a gradient of 17 : 1000 and 260 metres (285 yards) long it reached a speed of 6 km/h (3$^3/_4$ m.p.h.) with the same train and 60 km/h (37 m.p.h.) when running light.

DIESEL-ELECTRIC LOCOMOTIVES OF THE
CIE. FERMIERE DES CHEMINS DE FER TUNISIENS.

The locomotive shown in fig. 92 is built by the "Diesel-Elektriska Vagn Aktie-bolaget Västerås" in Sweden.

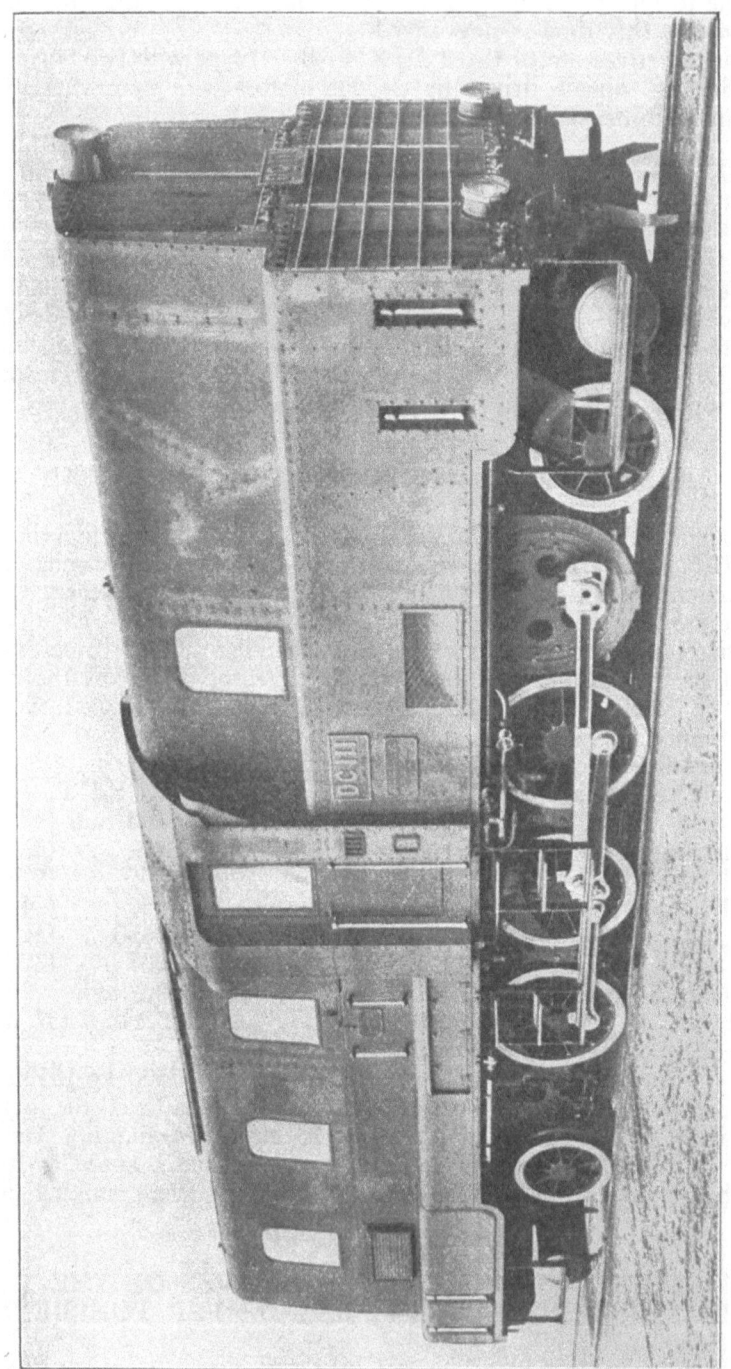

Fig. 91. Diesel-electric locomotive of the Japanese State Railways.

The main dimensions are:

Length over buffers	10.54 m (34' 7″)
Total wheel base	7 m (22' 11½″)
Distance centre to centre of bogies	5 m (16' 4⁷/₈″)
Bogie wheel base	2 m (6' 6³/₄″)
Total weight in working order	30 tons
Oil storage capacity	360 litres (80 glns.)
Water ,, ,, ...	260 ,, (57 ,,)

Engine
- cylinders 6
- bore. 200 mm (7⁷/₈″)
- stroke. . . . . 240 ,, (9⁷/₁₆″)
- rating . . . . 120 H.P.

The Diesel engine is a single-acting, four-stroke engine. The six cylinders are arranged in V form, so that the crank shaft has only three cranks.

The fuel is injected into the cylinders by means of air supplied from a compressor which is driven by the Diesel engine and which at the same time fills a vessel from which compressed air is drawn for injecting the fuel when starting up. Starting is actually done electrically. The fuel is pumped into the compressed air and the delivery is controlled by a governor according to the requirements of the Diesel engine.

The engine bearings are lubricated under pressure; the cylinders are not lubricated.

The cylinder cooling water is re-cooled in a cooler mounted on the roof.

The electric generator is coupled direct to the Diesel engine by a flexible coupling and serves as a motor for starting.

Fig. 92. Diesel-electric locomotive Tunisian Railways.

Each axle of the locomotive is driven by a motor via gear wheel transmission.

A driver's cabin is placed at each end of the locomotive.

Sufficient fuel can be carried for a distance of 800 km (500 miles) and the fuel consumption in normal service is 7 g per tkm including the locomotive and 13 g per tkm excluding the locomotive, on lines with gradients of 0—20 : 1000; the gauge is 1 metre (3'3³/₈″). Compared with steam locomotives the saving in fuel is about 50%.

In view of the results obtained with this locomotive an order was placed in 1924 with the "Cie. de Construction Mécaniques" of Paris for a second locomotive with a Diesel engine (Sulzer) of 250 H.P., likewise with two four wheeled bogies. The mechanical part was made by the "Cie. Francaise de Matériel de chemins de fer" and the electrical part by the "Cie. Electro-Mécanique" in cooperation with Brown Boveri & Cie.

The main dimensions of this second locomotive are:

Length over buffers . . . . 9.36 m (30′ 8½″)
Total wheel base . . . . 6.15 „ (20′ 2⅛″)
Distance centre to centre of bogies . . . . 4.25 „ (13′ 11¼″)
Wheel base of bogie . . . . 1.90 „
Total weight in working order . . . . 38.7 tons
Oil storage capacity . . . . 900 litres (200 glns.)
Water „ „ . . . . 1000 „ (220 „)

Engine
$\left\{\begin{array}{l} \text{cylinders} \\ \text{bore} \\ \text{stroke} \\ \text{revs./min} \\ \text{rating} \end{array}\right.$

cylinders 8
bore. . . . . 215 mm (8⁷⁄₁₆″)
stroke . . . . 300 mm (11¹³⁄₁₆″)
revs./min 550
rating . . . . 250 H.P.

Fig. 93. Train with Diesel-electric locomotive in Tunis.

This Diesel engine is likewise a single-acting, four-stroke engine with the cylinders placed in V form, but it is compressorless. The fuel delivery of the injection pump is controlled by a governor.

The cooling water is re-cooled in a tubular cooler divided into six parts arranged around an electrically-driven ventilator. One or more of the divisions can be disconnected in order to obtain the desired temperature.

The electric generator, coupled direct to the Diesel engine, supplies the current for the four traction motors driving the locomotive axles. The coupling· is on the Ward Léonard system, with a tension varying between 380 and 750 volts. The exciter (26 kW, 150 volts) is placed on top of the generator and is driven by means of a belt.

The axles of the locomotive are driven by the traction motors by means of a gear wheel transmission in the ratio of 16 : 71.

This locomotive was required to haul on the level a train of 80 tons at 60 km/h (37 m.p.h.),

on a gradient of 5 : 1000 at 37 km/h (23 miles per hour)
,, ,, ,, ,, 10 : 1000 ,, 26 ,, (16 ,, ,, ,,)
,, ,, ,, ,, 15 : 1000 ,, 19 ,, (12 ,, ,, ,,)
and in the trial runs it was found to meet these requirements.

The fuel consumption in normal service is about
7.2 g per tkm (0.025 lb./tonmile) incl. locomotive
and 9.8 ,, ,, ,, (0.034 lb./tonmile) excl. locomotive

Fig. 93 shows this locomotive coupled to a train at Tunis.

AMERICAN DIESEL LOCOMOTIVES.

In the United States the design of a Diesel locomotive has been taken in hand by the American Locomotive Co. in cooperation with the Ingersoll Rand Co. and the General Electric Co.

The combustion engine is a vertical, compressorless, single-acting, four-stroke engine with six cylinders, which can be run at various speeds. There is one fuel pump serving all cylinders, the fuel being distributed to the various cylinders by a distribution which allows the oil to pass to the burners just at the right moment. The oil for such an engine must be a pure hydrocarbon oil with a viscosity at 70° F. (21° C.) not exceeding 150 secs. Saybolt and a flashpoint not lower than 150° F. (66° C.).

All moving parts are lubricated under pressure. Oil which has been in contact with the cylinder walls is pressed through a filter and returned to the carter oil reservoir.

The cooling water is circulated by a centrifugal pump driven from the crank shaft. The temperature of the cooling water is controlled by a thermostat.

The generator is a 600 volt DC compound unit and is coupled direct to the combustion engine. Its voltage is adjusted to the requirements of the traction motors.

On the same shaft as that carrying the main generator there is a 60 volt exciter for exciting the field windings of the generator and a 32 volts accumulator battery. The exciter and accu battery current are automatically adjusted by a switch on the main control handle.

The traction motors are mounted in the bogies and drive the axles via gear wheels.

The first 60 ton locomotive has been in daily service in the goods yards of various railways and factories in the East of the States and has given some important economic results. In one goods yard the locomotive was used on shunting work for 68 days, most of the time for 24 hours per day (three shifts), being merely inspected when the drivers changed over.

From 9th June 1924 to 9th June 1925 the same locomotive was in service for 2217 hours, during which time 5358 miles (8650 km) were covered, with an average fuel consumption of one gallon per 7 kWh (one litre per 1.55 kWh). The following main values were found during this period:

Load factor. 16.6%
Cost of fuel { per running hour . . . . 34.3 ¢
{ per kWh . . . . 1.41 ¢

In the subjoined table some further particulars are given regarding the performance of the locomotive:

Class of service	Number of hours in service	Kilowatt hours generated	Loading factor in %	Consumption fuel oil in glns.	Lubricating oil in glns.	Coolingwater in glns.	Aux. petrol in glns.	Ton miles	Locomotive miles	Locomotive hours service	Expenses per kWh in dollars	Expenses per loc. mile in dollars	Fuel cost: Fuel oil at $ 0.05 per gln. Lub. oil at $ 0.50 " " Water at $ 0.13 per 1000 glns. Petrol at $ 0.17 per gln.
1. Shunting from July 10-August 23, 1924	833	15063	13.0	2400	249	508	30	400000	1531	0.297	0.0164	0.162	
2. Shunting from July 24-August 7, 1924	280	4098	12.05	672	80	—	—	112930	466	0.265	0.018	0.158	24 hours per day
3. Local goods service August 29, 1924	10	520	28.2	54.5	4	—	—	16615	47	0.472	0.0091	0.10	
4. Light yard drilling Sept. 5—9, 1924	35	499	8.7	80.75	12	—	—	—	42	0.287	0.020	0.239	
5. Shunting from Sept. 22-Nov. 4 1924	271	9385	21.1	1197.5	55	200	10	—	917	0.327	0.0094	0.097	
6. Shunting Nov. 28-Dec. 7, 1924	128	4630	20.6	592.5	30	—	—	—	318	0.349	0.0096	0.140	
7. Passenger service Feb. 27—28 1925	11.12	678	30.5	88	10	—	—	28500	194	0.840	0.0138	0.048	2 passenger coaches Jersey-City—Harrisburg
8. Tonnage test Aug. 14, 1924	0.36	65	91.0	6	0.154 est.	—	—	4125	1	1.05	0.0058	0.377	93 loaded freight cars on level track.
9. Shunting Mar. 4-17, 1925	207	6383	18.1	945.5	89	200	6	—	529	0.45	0.014	0.175	
10. Shunting Mar. 23-28, 1925	120	4180	19.8	621.5	50	50	—	—	329	0.467	0.013	0.170	
11. Shunting July 9-11, 1925	32	1505	27.5	175	20	—	—	—	120	0.58	0.012	0.156	
12. Total & Average 1 year's operation	2217	58451	16.6	8440.5	711	1615	46	—	5358	0.343	0.0141	—	

In order to determine the wear of the various parts of the combustion engine inspections were made on 12th February and 12th September 1924 (six months later). The second inspection was made in the workshops of the Ingersoll Rand Co. at Philipsburg, New Jersey, and as a result it was found that after six months' service practically no wear had taken place. The maximum wear of the crosshead pins was 0.008″ (0.02 mm).

The main dimensions of these locomotives are given below.

A. *Shunting locomotive.*   *Metric measure*

Weight	60 tons	54 tons
rating	300 H.P.	300 H.P.
max. speed	30 m.p.h.	48 km/h
tractive force at start	36000 lbs.	16000 kg
„ „ at 10 m.p.h.	8000 lbs.	3600 kg

B. *Train locomotive.*

Weight	60 tons	54 tons
rating	300 H.P.	300 H.P.
max. speed	55 m.p.h.	90 km/h
tractive force at start	23000 lbs.	10000 kg
„ „ at 21 m.p.h.	4000 lbs.	1800 kg

C. *Shunting locomotive.*

Weight	100 tons	90 tons
rating	600 H.P.	600 H.P.
max. speed	30 m.p.h.	48 km/h
tractive force at start	60000 lbs.	27000 kg
„ „ at 10 m.p.h.	15500 lbs.	7000 kg

D. *Train locomotive.*

Weight	100 tons	90 tons
rating	600 H.P.	600 H.P.
max. speed	55 m.p.h.	90 km/h
tractive force at start	54000 lbs.	24000 kg
„ „ at 25 m.p.h.	6200 lbs.	2750 kg

E. *Shunting locomotive.*

Weight	108 tons	97 tons
rating	750 H.P.	750 H.P.
max. speed	30 m.p.h.	48 km/h
tractive force at start	64800 lbs.	29000 kg
„ „ at 10 m.p.h.	20000 lbs.	9000 kg

F. *Train locomotive.*

Weight	108 tons	97 tons
rating	750 H.P.	750 H.P.
max. speed	55 m.p.h.	90 km/h
tractive force at start	58400 lbs.	26000 kg
„ „ at 25 m.p.h.	7800 lbs.	3500 kg

WHITCOMB LOCOMOTIVES.

The Geo. D. Whitcomb Company of Rochelle, Ill. U.S.A., build motor-locomotives in several standard types.

Fig. 94. Chain drive of the Whitcomb locomotives.

Transmission is by means of chains, two to one driving axle and another chain from the first axle to the other (see fig. 94). This avoids a one-sided drive and the driving couple acts equally on all driving wheels.

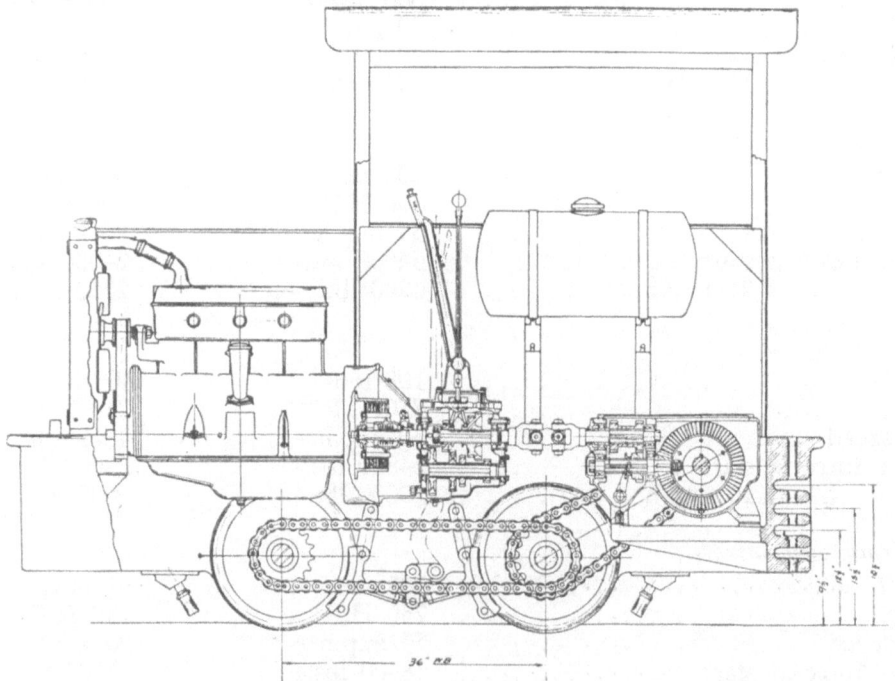

Fig. 95. Whitcomb Petrol Locomotive.

The speeds are varied by means of gear wheel transmission contained in a closed box. The reversing gear is in a separate box.

The engine is started electrically by an electric motor fed by current from an accumulator battery. Once the engine is running the electric motor serves as generator, supplying current for charging the accumulator battery, from which current is also drawn for the electric head lights.

The shafts rotate in Hyatt roller bearings.

In most of the places where these locomotives are used there is no opportunity to overhaul or carry out repairs to the locomotive from a repair-pit, and the parts

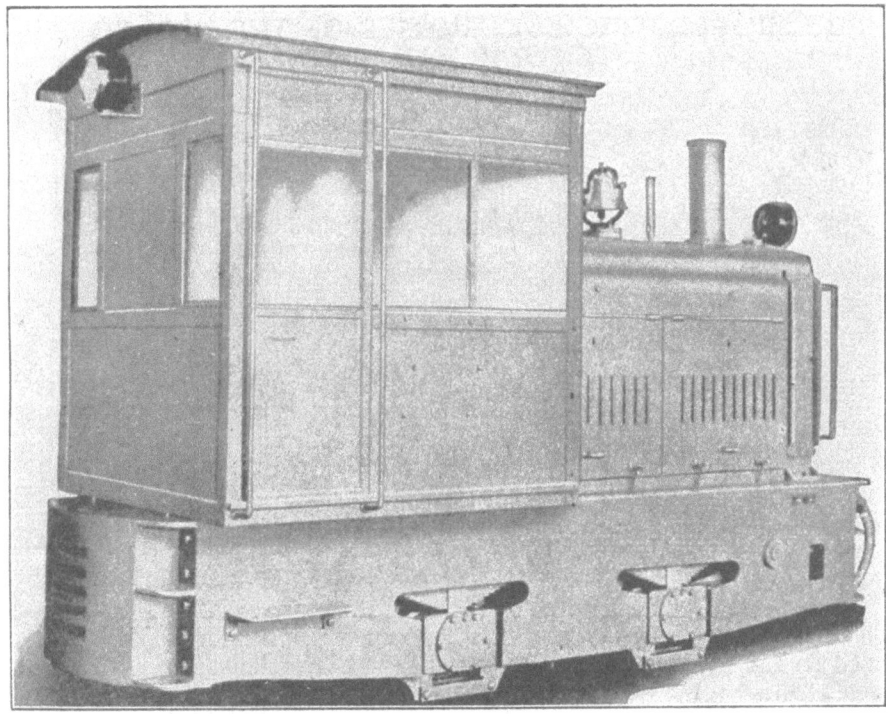

Fig. 96. Whitcomb Diesel locomotive.

subject to wear are therefore arranged in such a way that they can easily be reached without another part having first to be removed.

These locomotives are built for various gauges and in various sizes, with petrol and with Diesel engines.

Fordson tractor engines are used for petrol power and two-cycle Fairbanks-Morse compressorless engines as Diesel engines.

Petrol locomotives are built in sizes weighing 2½, 3½, 4½, 5, 6, 7, 8, 12, 14, 16, 20 and 25 tons, and Diesel locomotives in weights of 6, 8, 12 and 20 tons. Fig. 92 shows a petrol locomotive and fig. 93 a Diesel locomotive. The main dimensions are given of two of these locomotives.

	5 ton Petrol loc.	20 ton Diesel loc.
Length	112″ (2.845 m)	17′ 4″ (5.283 m)
Wheel base	36″ (914 mm)	52″ (1321 mm)
Wheel diameter	18″ (457 mm)	24″ (610 mm)
Cylinders	4	6
Bore	4¼″ (108 mm)	6″ (152 mm)
Stroke	4½″ (114 mm)	6½″ (165 mm)
Engine power	40 H.P.	180 H.P.
Max. speed	12.6 m.p.h. (20 km/h)	14 m.p.h. (23 km/h).

DIESEL-ELECTRIC LOCOMOTIVE OF THE NEW YORK CENTRAL RAILROAD.

Fig. 97. Diesel-electric locomotive of the New York Central Railroad.

This locomotive is in service in the Putnam division in New York State. It is illustrated in fig. 97 and its main dimensions are as follows:

Type.	2 D_0 2 (4—8—4)
Power	750 H.P.
Engine revs./min	240/520
Cylinders per engine	6
Number of engines	4
Generator tension	600 volts DC
Dia. of driving wheels	44″ (1118 mm)
Rigid wheel base	17′ 6″ (5.334 m)
Total ,, ,, ...	42′ 10″ (13.055 m)
Length overall	51′ 7″ (15.722 m)
Length of body	46′ 5½″ (14.161 m)
Height over radiator	14′ 9½″ (4.509 ,,)
Greatest width	10′ 0″ (3.018 ,,)
Total weight	300,000 lbs. (135 tons)
Adhesion weight	170,000 ,, (76 tons)

Tractive force at start . . . .   51,000 lbs. (22,500 kg)
„   „   „ 5 m.p.h. (8   km/h) . . . .   30,000 „   (13,400 „)
„   „   „ 20 „   (32   „) . . . .   10,000 „   (4,500   „)
„   „   „ 40 „   (64   „) . . . . . .   4,000 „   (1,800   „)

This locomotive was built by the American Locomotive Co. in cooperation with the General Electric Co. and the Ingersoll Rand Co.

Fig. 98. Diesel-electric locomotive of the Canadian National Railways.

DIESEL-ELECTRIC LOCOMOTIVE OF THE CANADIAN NATIONAL RAILWAYS.

The Canadian National Railways have taken a Diesel electric locomotive (No. 9000) into service with which trials have been made in comparison with a steam locomotive (No. 6000).

The Diesel-electric locomotive consists of two units, each with a 12 cylinder compressorless Diesel engine coupled direct to a Westinghouse generator supplying current to four traction motors each driving an axle. These Diesel engines have a power of 1330 H.P. each at 800 revs./min. The engine speed can be reduced to 300 revs./min.

This locomotive is illustrated in figs. 98 and 99 and its main dimensions are:

Total weight . . . . . 290 tons
Adhesion weight . . . 214 tons
Driving axles 8
Dia. of driving wheels 51″ (1295 mm)
Transmission ratio. . . 22 : 69
Max. tractive force . . 100,000 lbs. (45000 kg)
Make of Diesel engines: Beardmore
Fuel oil capacity . . . 2 × 8000 lbs. (2 × 3600 kg)
Lub. oil   „   . . . 2 × 1000 „ (2 × 450 kg)
Cooling water capacity . 2 × 3000 „ (2 × 1360 kg)

The fuel consumption is:

Revolutions	Output H.P.	Fuel cons. per H.P.h lbs.	grammes
800	1360	0.41	185
700	1190	0.395	180
600	1000	0.39	179
500	840	0.40	182
400	1530 (overloaded)	0.43	196

A trial run was made on 26th August 1929 from Montreal to Toronto with a train running to the same schedule as fixed for the "International Limited" train drawn by a steam locomotive.

The steam locomotive (No. 6000) is of the 2 D 1 (4—8—2) type with twelve-wheeled tender and its main dimensions are the following:

Total weight . . . . 595 tons
Adhesion weight . . . . 231 „
Driving axles 4
Dia. of driving wheels . . . . 73″ (1854 mm)
Heating surface of boiler . . . . 4049 sq. ft. (375 m²)
Superheater surface . . . . 810 „ „ (75 „)
Grate area . . . . 66.7 „ „ (6.20 „)
Cylinders . . . . 26″ × 30″ (660 × 762 mm)
Steam pressure . . . . 210 lbs./sq. in. (14.6 kg/cm²)
Max. tractive force . . . . 49,600 lbs. (22,000 kg)

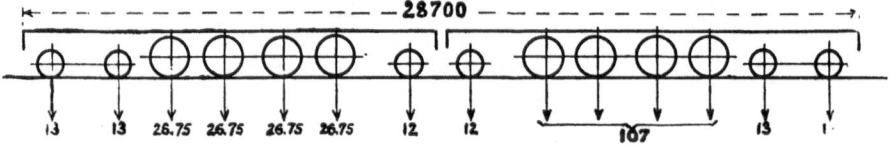

Fig. 99. Diagram of axle load of the Diesel-electric locomotive Canadian National Railways.

The following are particulars of the trial run with the Diesel electric locomotive:
Distance Montreal Toronto . . . . 334 miles (540 km)
Number of coaches 7
Train weight excl. locomotive . . . . 663 tons
Time taken for the run . . . . 7 hours 40 min
Number of stops 13
Average speed . . . . 43.6 m.p.h. (70 km/h)
Power:
at speed of 60 m.p.h. (96 km/h) max. . . . . . . 2100 H.P.
at speed of 45 „ (72 „) on gradients of sev-
 eral minutes' run . . . . 2500—2600 H.P.
at max. speed of 73 m.p.h. (118 km/h) . . . . . . 1800 H.P.
 The normal speed is 60—65 m.p.h. (96—103 km/h).
 The cost for fuel was about one-third of that for the above-mentioned steam
locomotive drawing the same train.

DIESEL-ELECTRIC LOCOMOTIVE OF THE CENTRAL RAILROAD OF NEW JERSEY.

This locomotive (fig. 100) was taken into service in October 1925. It has three
compartments, the middle one containing the machinery and the two end ones forming the driver's cabins. The load of the machinery is distributed evenly over all
axles.
 Compressed air of 200 lbs./sq.in. (14 kg/cm²) is used for starting.
 The locomotive was put into service at the Bronx Terminal, New York City
and was the first Diesel-electric locomotive used on the railways in the U.S.A. It
has given very satisfactory results and from December 1925 to March 1926 trial
runs were made with it for comparison with runs made by a steam locomotive in
the same months of the previous years, with the results given below:

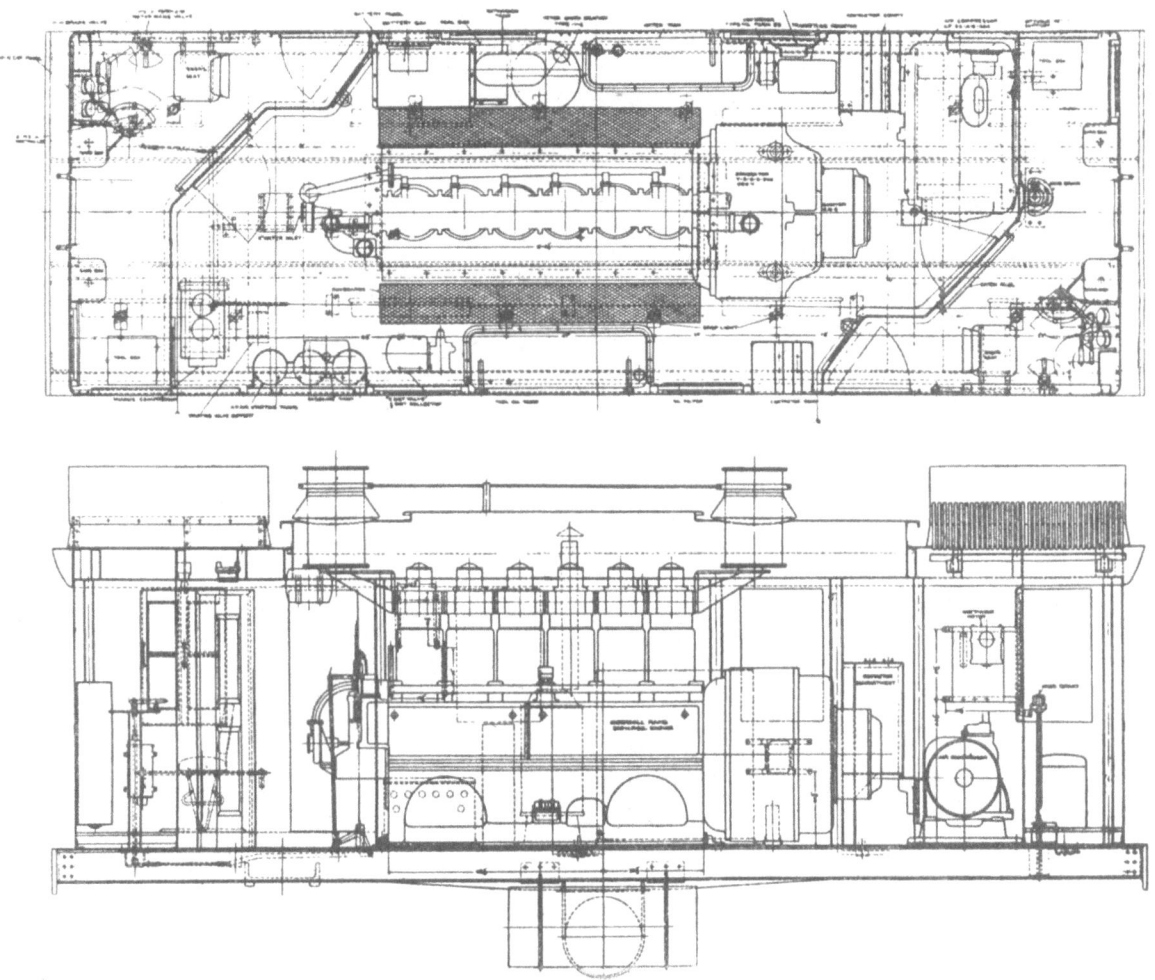

Fig. 100. Interior of 60 ton oil-electric locomotive, Central Railroad of New Jersey.

	Diesel electric locomotive	Steam locomotive
Number of service days	101	100
„ „ „ hours	1530	1207
„ „ fuel consuming hours	1433	1207
Fuel consumption in gallons	4339	—
Lubricating oil in gallons	185	40
Water in gallons	896	—
Kilowatthours	26465	—
Coal consumption in tons	—	169
Number of floats in	327	176
„ „ „ out	327	174
Total number of floats	654	350
Number of cars off float	3583	2901
„ „ „ on „	3570	2648

	Diesel electric locomotive	Steam locomotive
Total number of cars	7153	5749
Tons handled off floats, weight of cars included	148704	118075
„ „ on „ „ „ „ „	85717	72666
Total tons handled „ „ „ „ „	234421	190741
Total cost of fuel, lubricating oil and water . . .	$ 300.76	$ 1420.13
Average number of service hours per day . . .	15.1	12.07
Fuel consuming hours per day	14.2	12.07
Fuel consumption per day in gallons	42.9	
Coal consumption per day in tons		1.69

The following unit prices were taken for the calculations:
fuel oil 5 ¢ per gallon; Diesel lubricating oil 53—54 ¢ per gallon; water $ 1.00 per 1000 cub.ft.; coal $ 7.15 per ton.

The main dimensions are given in the table below:

Engine:

Type: Ingersoll-Rand, compressorless, 4 cycle, vertical.

Number of cylinders	6
Cylinder bore	10″ (254 mm)
Stroke	12″ (305 mm)
Revs/min.	600
Output	300 H.P.
Piston speed	1200 ft. p. min (6 m/sec)
Fuel oil consumption	0.43 lb. per brake horse-power.
Cooling surface of radiator	1200 sq.ft. (110 m²)

Fuel distribution by rotating distribution valves.
Lubrication under pressure with filtration.

Generator:

Type:	200 kW—600 volts—600 revs./min
Exciter	6 kW—direct coupled—60 volts
Shunt windings:	separately excited
Serie „ :	differential compound
Voltage :	200—750 volts.

Tractions motors:

Number	4
Nominal output	95 H.P. each at 600 volts.
Connection	2 in parallel, series and parallel grouping.
Transmission	gear wheel 82 teeth
	pinion 14 „

Locomotive:

Length	32′ 8″ ((9.957 m)
Width of the body	9′ 4″ (2.845 m)
Max. width	10′ 0″ (3.048 m)
Rigid wheel base	7′ 2″ (2.184 m)
Total weight	120,000 lbs. (55 t)
Adhesion weight	120,000 lbs. (55 t)
Weight on each axle	30,000 lbs. (13.7 t)
Tractive force	36,000 lbs. (16400 kg) at an adhesion coëf. of 30% up to about 1 m.p.h. (1.6 km/h).

Fig. 101. 60 Ton oil-electric locomotive, Central Railroad of New Jersey.

Fig. 102. 60 Ton oil-electric locomotive, Baltimore & Ohio Railroad.

The particulars of the 60 ton train locomotive are the same, except for the transmission ratio.

Fig. 101 is an illustration of this locomotive and fig. 102 shows a similar one of the Baltimore and Ohio Railroad.

DIESEL-ELECTRIC LOCOMOTIVE OF THE LONG ISLAND RAILROAD COMPANY.

The main dimensions of this locomotive are:

Engines:

Number 2
Type: Ingersoll-Rand, 4 cycle, vertical, compressorless.
Number of cylinders per engine . . 6
Bore . . . . 10″ (254 mm)
Stroke . . . . 12″ (305 mm)
Revs./min 600
Total output of both engines. . . . 600 H.P.
Piston speed . . . . 1200 ft. p. min (6 m/sec)
Fuel oil consumption . . . . 0.43 lbs. (190 g) per brake horse-power hour
Cooling surface of radiator . . . . 2400 sq.ft. (230 m²)
Fuel distribution, by rotating distributing valves.
Lubrication under pressure with filtration.

Generators.

Number 2
Type. . . . . 200 kW, 600 volt, 600 revs. p. min
Exciter . . . . 6 ,, direct coupled, 60 volts
Field windings . . . . separately excited
Serie winding . . . . differential compound
Voltage . . . . 200—750 volts

Traction motors:

Number 4
Normal output . . . . 200 H.P. each at 600 volts.
Connection . . . . 2 in parallel, series and parallel groupings
Transmission . . . . gear wheel 70 teeth
                                    pinion   15   ,,

Compressors:

Number 4
of which one for the air brake, with a piston displacement of 100 cu.ft. (2.8 m³) per min; pressure 90—140 lbs. per sq.in. (6—10 kg/cm²); one mechanically driven by each engine for starting;
and one driven by an auxiliary engine, also for delivering air for starting.

Fig. 103. Bogie for 100 ton oil-electric locomotive, Long Island Railroad Co.

Locomotive:
Length   48′ 2″ (14.681 m)
Width of body
          10′ 0″ (3.018 m)
Rigid wheel base.
          7′ 2″ (2.184 m)
Total weight
          200,000 lbs. (90 t)
Adhesion weight
          200,000 lbs. (90 t)
Weight on each axle
          50,000 lbs. (22 t)
Tractive force
          60,000 lbs. (26500
          kg), at an adhe-
          sion factor of 30%
          up to about 1 m.
          p.h. (1.6 km/h)

Except for the necessary modifications in the transmission the particulars of the 100 ton train locomotive are practically the same as those given above.

Figs. 103 and 104 give an idea of the bogie and the control stand of this locomotive.

A trial run of 537 miles (860 km) was made with the locomotive described above on 16th December 1925 fron Erie, Pa., to Greenville, N. J. (see fig. 105). The run was over the Pensylvanian Railroad via Williamsport, Harrisburg and Trenton Junction with a train of 5 loaded cars, a passenger coach and luggage van, with a total weight (incl. the locomotive) of 377 tons. The run took in all 40 hours 24 min, of which $28^3/_4$ running hours. The total consumption of fuel oil was 473 glns. (2150 litres), which works out at an average consumption of

6.35 lbs. per loc.mile (1.8 kg per loc. km). The lubricating oil consumption was 5 glns. (22^1/$_2$ litres).

The cost of oil was: fuel oil 473 × 5 ¢ = $ 23.65
lub. „ 5 × 50 ¢ = $ 2.50
—————
$ 26.15

for 202,449 ton miles, against which 3810 kWh were generated.

Fig. 104. Driver's cabin 100 Ton oil-electric locomotive, Long Island Railroad Co.

The locomotive is built for a max. speed of 30 m.p.h. (48 km/h) and for shunting service. The average speed on the above trial run was 18.7 m.p.h. (30 km/h) and the steepest gradient was 16 : 1000 extending for a distance of about 8 miles (13 km).

Fig. 105. Oil-electric locomotive leaving Erie, Dec. 15, 1925, Long Island Railroad Co.

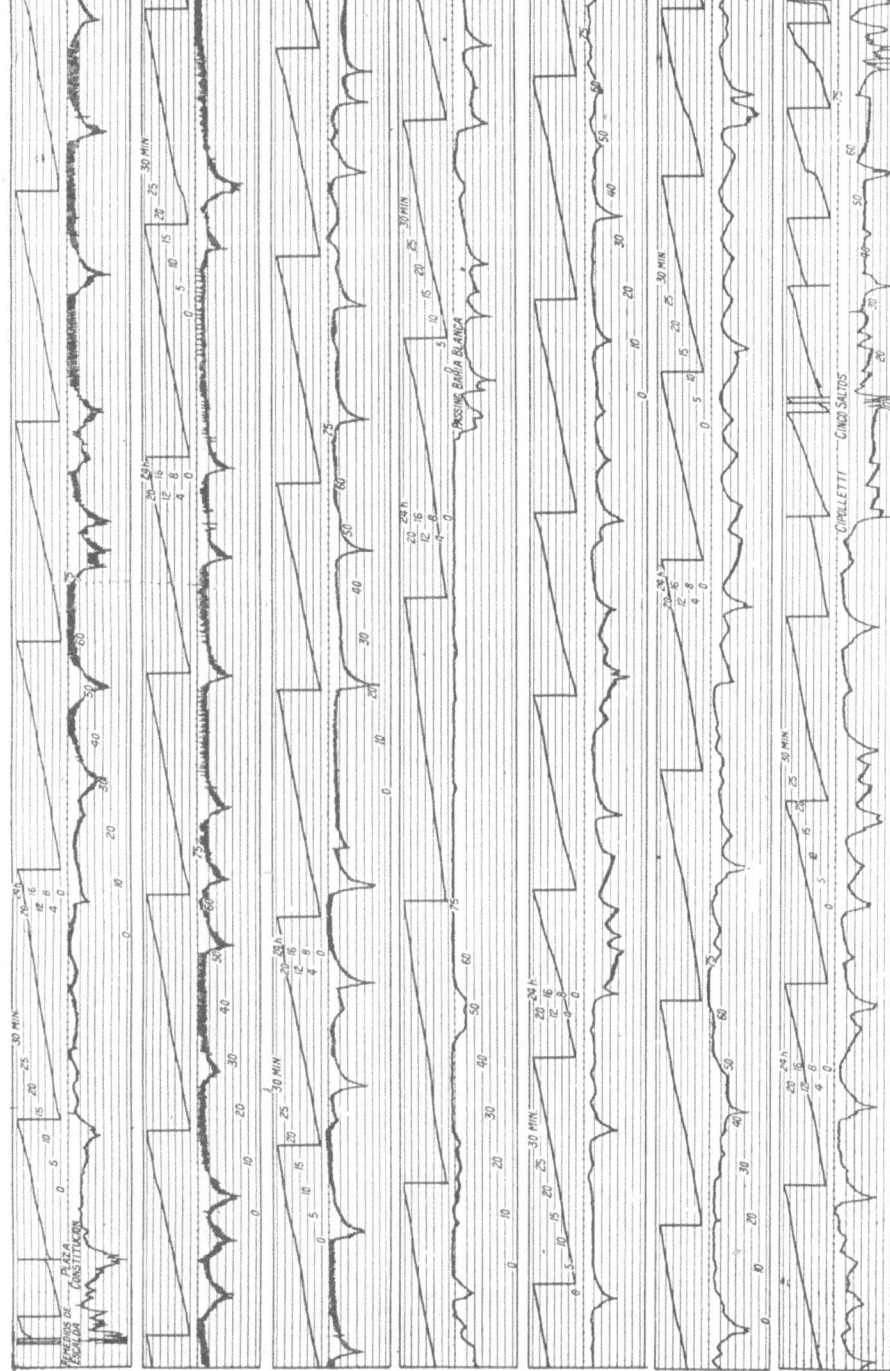

Fig. 106. Speed and time diagram of trial run of Diesel-electric locomotive, Buenos Aires Great Southern Railway.

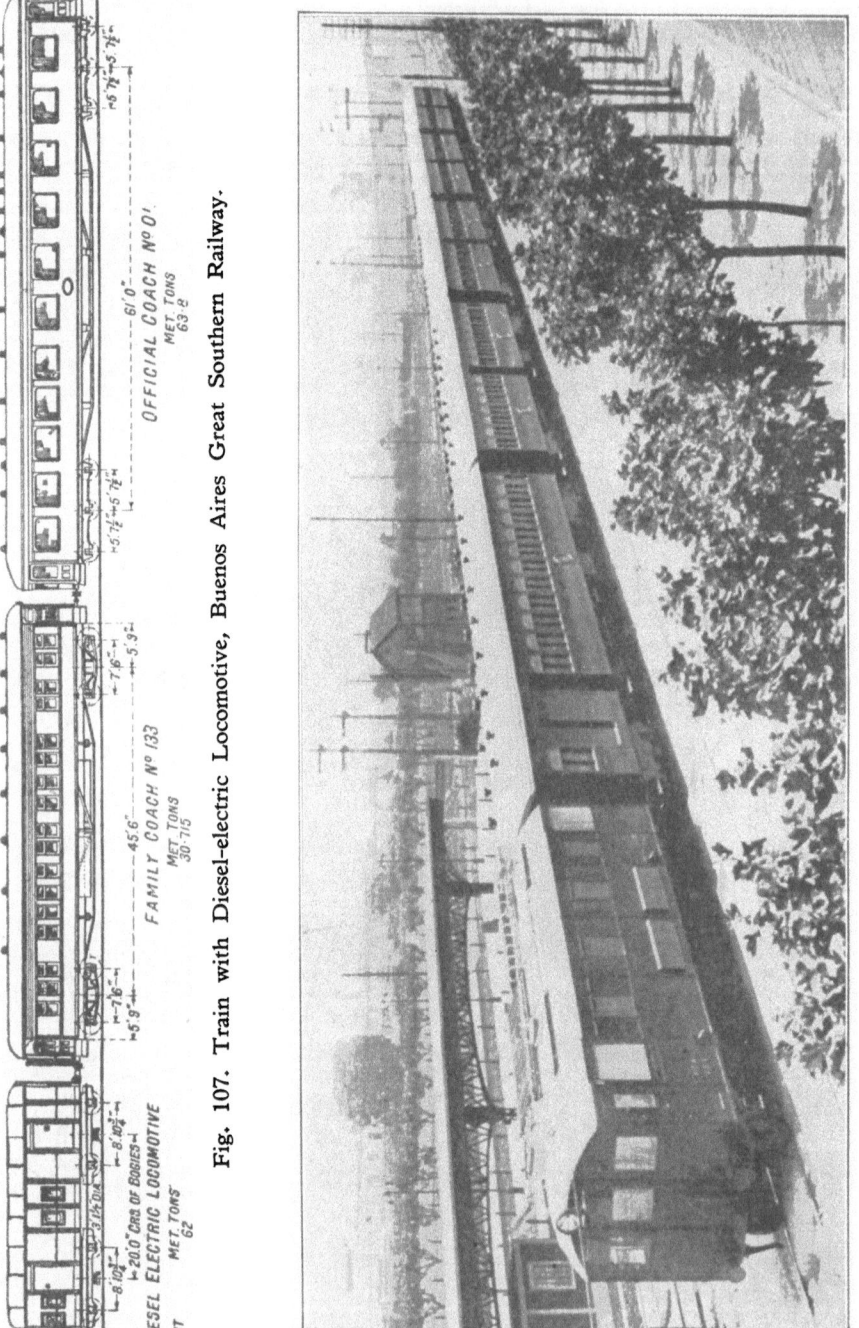

Fig. 107. Train with Diesel-electric Locomotive, Buenos Aires Great Southern Railway.

Fig. 108. Train with Diesel-electric locomotive, Buenos Aires Great Southern Railway.

$B_0 + B_0$ (0—4—0 + 0—4—0) DIESEL-ELECTRIC LOCOMOTIVE OF THE BUENOS AIRES GREAT SOUTHERN RAILWAY [1].

This locomotive was built with a Beardmore 8-cylinder, four-cycle, Diesel engine with a normal brake power of 375 H.P. at 700 revs./min. The engine is coupled direct to a Metropolitan-Vickers compound DC generator of 300 kW, 750 volts.

A trial non-stop run was made over a distance of 770 miles (1240 km) with a train of 159 tons in 20 hours 37 min at an average speed of 37 m.p.h. (60.1 km/h). The total fuel consumption was 2170 lbs. (986 kg), that is 2.8 lbs. per mile (0.79 kg/km). The calorific value of the fuel was 10550 kcal/kg. The cooling water temperature was kept at 65° C. (149° F.) in spite of the hot and sunny weather and the coolers being exposed on the roof; at night the atmospheric temperature dropped to below freezing point. The electric generator supplied 2980 kWh, i.e. 24 Wh per ton mile (15 Wh/tkm).

The run was from Buenos Aires to Cipolletti with a four-axled coach (31 tons) and a six-axled coach (64 tons), whilst the weight of the locomotive was 62 tons, making in all 156 tons with 3 tons load. The max. speed of 47 m.p.h. (75 km/h) was maintained over a distance of 45 miles (73 km).

Fig. 106 gives an idea of the trial run and fig. 107 shows the train as composed for the trial run. In fig. 108 the Diesel-electric locomotive is seen coupled to an ordinary passenger train [2].

---

[1] The Railway Gazette, Aug. 16, 1929, p. 255.
[2] The Railway Gazette, March 14, 1930, p. 394.

MAIN DIMENSIONS OF LOCOMOTIVES.

D = Diesel engine   d = direct   h = hydraulic   p = pneumatic
E = Explosion engine   e = electrical   m = mechanical

Railways	Gauge ft.	Gauge metres	Locomotive type	Kind of engine	Engine rating H.P.	Motion transmission	Max. speed km/h	Max. speed m.p.h.	Weight in working order in tons
Russia	5′	1.524	1 E_0 1 (2-10-2)	D	1200	e	55	34	120
,, ...	5′	1.524	2 E 1 (4-10-2)	D	1100	m	50	31	131
,, ...	5′	1.524	1 E_0 1 (2-10-2)	D	1200	e	60	37	135
Austrian Federal Railways	4′ 8½″	1.435	C_0 (0-6-0)	D	200	e	60	37	37
German State Railway ..	4′ 8½″	1.435	2 C 2 (4-6-4)	D	1200	p	80	50	120
Sweden...............	,,	,,	2 B 2 (4-4-4)	D	300	h	75	47	
Italian State Railways ..	,,	,,	1 C 1 (2-6-2)	D	200	h	37	23	40
,, ,, ,, ..	,,	,,	2 C 1 (4-6-2)	D	700				94
Swiss Federal Railways..	,,	,,	B (0-4-0)	E	100	m	15	9	21
Netherlands Railways....	,,	,,	B (0-4-0)	E	50	m	25	15	12
,, ,, ...	,,	,,	B_0 (0-4-0)	D	150	e	60	37	30
Danish State Railways ..	,,	,,	1 B_0 2 (2-4-4)	D	450	e	80	50	55
French State Railways ..	,,	,,	B (0-4-0)	E	100	m	60	37	30
Paris-Lyons Med. Railway	,,	,,	B (0-4-0)	E	40	m	33	21	17
French Northern	,,	,,	B (0-4-0)	E	55	m	16	10	15
,, ,, ...	,,	,,	B_0 (0-4-0)	E	90	e	25	15	20
French Eastern Railway..	,,	,,	B_0 (0-4-0)	E	40	m	40	25	22
Spanish Northern Railway	5′ 6″	1.680	1 A (2-2-0)	E	50	m	20	12	9
Siamese State Railways..	3′ 3″	1.000	B (0-4-0)	D	200	m	40	25	24
,, ,, ,, ..	,,	,,	2 D_0 + D_0 2 (4-8-0 + 0-8-4)	D	1400	e			128
,, ,, ,, ..	,,	,,	2 D_0 2 (4-8-4)	D	900	e	60	37	88
Tunis	,,	,,	B_0 + B_0 (0-4-0 + 0-4-0)	D	120	e			30
,, ...	,,	,,	B_0 + B_0 (0-4-0 + 0-4-0)	D	250	e	60	37	39
Japan	3′ 6″	1.067	1 C 1 (2-6-2)	D	600	e	60	37	58
Central R.R. of New Jersey	4′ 8½″	1.435	B_0 + B_0 (0-4-0 + 0-4-0)	D	300	e	48	30	63
Baltimore & Ohio	,,	,,	,, ,,	D	300	e	48	30	63
Lehigh Valley Railroad ..	,,	,,	,, ,,	D	300	e	48	30	63
	,,	,,	,, ,,	D	300	e			68
Long Island Railroad....	,,	,,	,, ,,	D	600	e	48	30	100
	,,	,,	,, ,,	D	600	e	40	25	78
Great Northern Railroad	,,	,,	,, ,,	D	600	e	48	30	100
Erie Railroad	,,	,,	,, ,,	D	300	e	48	30	63
	,,	,,	,, ,,	D	600	e	48	30	100

MAIN DIMENSIONS OF LOCOMOTIVES (continued).

Railways	Gauge		Locomotive type	Kind of engine	Engine rating H.P.	Motion transmission	Max. speed		Weight in working order in tons
	ft.	metres					km/h	m.p.h.	
Chicago & N.W.	4' 8½"	1.435	$B_0 + B_0$ (0-4-0 + 0-4-0)	D	300	e	48	30	63
Reading	,,	,,	,, ,,	D	300	e	48	30	63
Delaware Lackawana & Western	,,	,,	,, ,,	D	300	e	48	30	63
New York Central	,,	,,	,, ,,	D	300	e	48	30	116
,, ,, ,, ...	,,	,,	2 D_0 2 (4-8-4)	D	750	e	65	40	132
,, ,, ,, ...	,,	,,	,, ,,	D	900	e	96	60	162
Illinois Centr.	,,	,,	$B_0 + B_0$ (0-4-0 + 0-4-0)	D	300	e	48	30	63
Canadian Nat.	,,	,,	2 D_0 1 + 1 D_0 2 (4-8-2+2-8-4)	D	2660	e	96	60	290
Chicago Gr. West	,,	,,	C (0-6-0)	E	225	m	24	15	27
Chicago Burlington & Quincy	,,	,,	B_0 (0-4-0)	E	85	e	15	9	27
Boston & Maine	,,	,,	2 D 2 (4-8-4)	D	1300	m	96	60	145
Buenos Aires Great Southern	5' 6"	1.680	$B_0 + B_0$ (0-4-0 + 0-4-0)	D	375	e	75	47	62

IV. TYPES OF MOTOR COACHES ALREADY BUILT.

DIESEL MOTOR COACH OF THE HALBERSTADT—BLANKENBURG RAILWAY *).

In view of the steep gradients (60 : 1000) occurring on this railway it was necessary to make the motor coach of the lightest possible construction, a comparatively large part of the power being required to overcome the gradients and only a relatively small part for propulsion. As it was not intended to couple the motor coach to other carriages there was no need of the usual buffer and draw gear and consequently it was not necessary to design the frame for the absorption of pulling

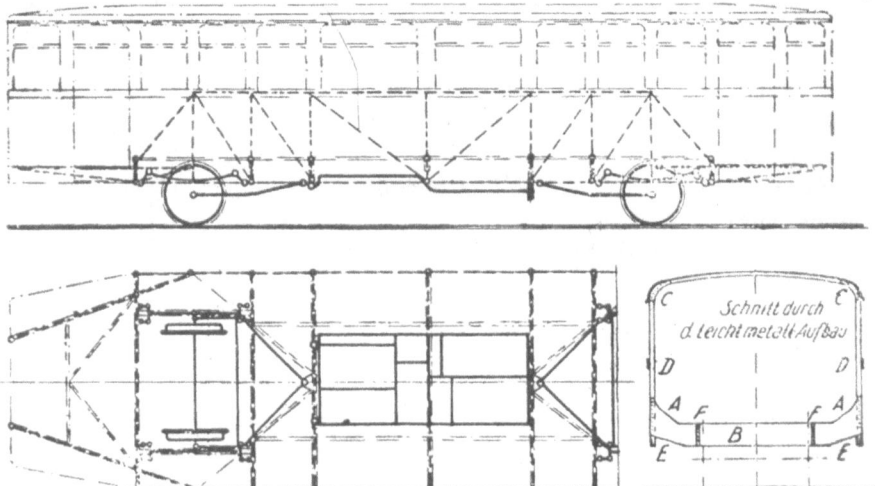

Fig. 109. Diesel motor coach Halberstadt—Blankenburg Railway (Germany). Frame.
Schnitt durch d. Leichtmetall Aufbau = Section through car body of light metal.

and pushing strains, so that all it had to bear was the weight of the coach body and the passenger load. This led to the sides of the body being made as bearing walls, in which the diagonals are replaced by sheet iron plates. Further, light metal has been used as far as possible, only those parts subject to strains that can hardly be taken up by light metal being made of iron or steel, although even in these parts the greatest attention was paid to the reducing of the weight wherever possible.

Fig. 109 shows the frame of this vehicle.

In order to facilitate the rounding of curves each of the two axles is built in a single-axle bogie as shown in fig. 110. This design was adopted because radial axles had only proved satisfactory for trailing coaches and not for motor coaches,

*) see: Sonderdruck aus Verkehrstechnik, 1928, Heft 38a. Ein neuer Leichtmetall-Diesel-Triebwagen mit mechanischer Kraftübertragung bei der Halberstadt-Blankenburger Eisenbahn von General-direktor Dr: Ing. e.h. Steinhoff und Regierungsbaumeister a.D. Dr. Kettler, Blankenburg (Harz).

experience having shown that the front axle does not adjust itself to the centre of the curve as is the case with a Bissel-bogie, the axle of which has to turn on the bogie-pivot, the parallel and sliding spring links bearing the load of the body always bringing about a return to the centre just as with radial axles. This has been proved in the trial runs, the coach running exceptionally smooth round curves. In order to reduce the weight of the non-elastic parts the wheel sets have been made in a special form (see fig. 111).

The wheel tyres are 25 mm thick and are bolted to a pressed steel disc, which in turn is bolted to a collar on the axle. Such a wheel set weighs about 600 kg, as against 1200 kg of an ordinary wheel set.

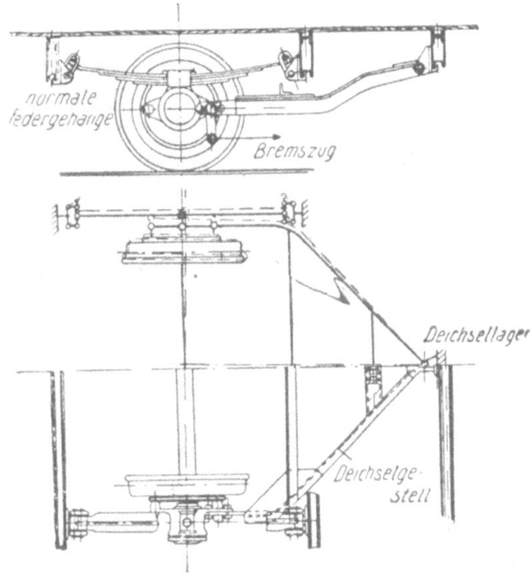

In view of the comparative thin-ness of the tyre there are no brake shoes acting on the tyre, but the braking is effected on the inside of a special brake rim (see figs. 110 and 111).

Fig. 110. Diese! motor coach Halberstadt—Blankenburg Railway (Germany).
Single-axle bogie.
Bremszug = Brake draw gear. — Deichselgestell = Single axle bogie frame. — Deichsellager = Pivot of Single axle-frame. — Normale federgehänge = Normal spring-suspension.

When a tyre has become worn it can be replaced by a new one without having to remove the wheel axle.

Fig. 109, from which may be seen which parts are made of light metal (—·— and - - -) and which

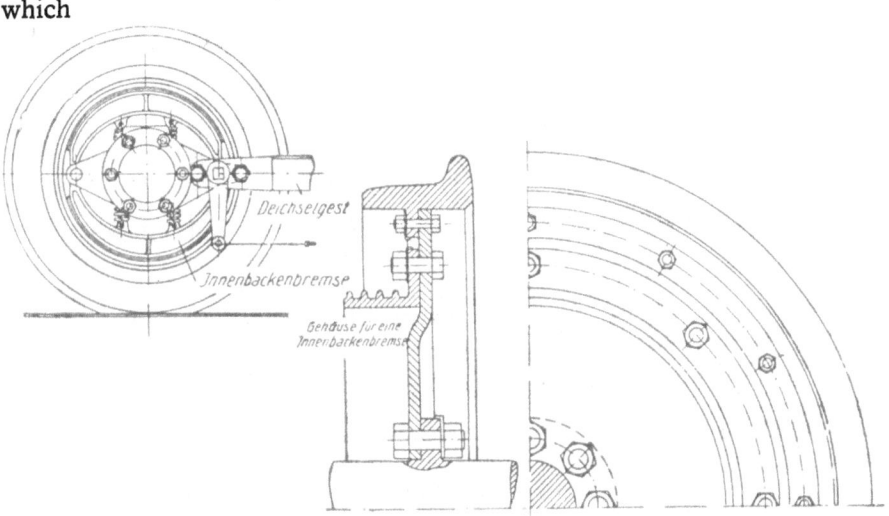

Fig. 111. Diesel motor coach Halberstadt—Blankenburg Railway (Germany). Wheel.
Deichsel gest = Single axle-bogie frame. — Innenbackenbremse = Internal expanding brake.
— Gehäuse für eine Innenbackenbremse = Brake drum for internal expanding brake.

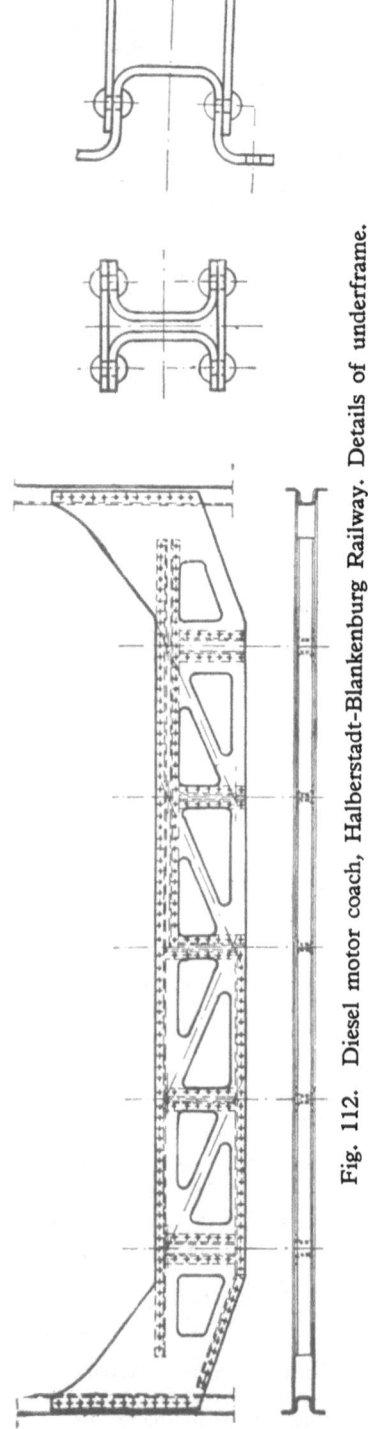

Fig. 112. Diesel motor coach, Halberstadt-Blankenburg Railway. Details of underframe.

of steel (———), shows the position of the axles in the Bissel-bogies and the frame for the machinery between the cross members of the coach body. The horizontal stresses acting on the wheel axles in a longitudinal direction are taken up by steel parts and without any special frame being used these stresses are not transmitted to the light metal body. The light metal used is "Lautal", an alluminium alloy of the "Vereinigte Aluminium-Werke" (Lauta-Werk). When this metal is used the following requirements have to be borne in mind, in view of its peculiar physical properties:

1) the plates should be as thin as possible, as otherwise the desired high ultimate strength is not reached;
2) rivetting is to be done cold, with rivets of small diameter (at most 12 mm), because larger rivets do not possess the necessary shearing strength.

In connection with the above the construction is made as follows:

The cross members, which consist of two plates provided with the necessary recesses, are kept apart by a bottom and top curb, whilst at the points where compression strains occur the two plates are joined together by plates in bent channel shape. These cross members are fixed direct to the uprights of the coach body (see fig. 112). These parts are connected to the roof braces, which are made of channel iron, by means of gusset plates so that the cross members, posts and roof braces together form one closed truss. These trusses are connected up in longitudinal direction by the roof curb at C, a reinforcing ledge D underneath the window frames, a reinforcement E at the bottom and finally by longitudinal members F between the cross members, as shown in figs. 113—115.

The engine is built in a special frame suspended from the cross members, and in order to avoid transmission of the engine vibration to the body the engine frame is hinged on one side at two points and on the engine side suspended by a spring at one point. This frame with engine and transmission gear can easily be removed. The top of the engine protrudes into the coach body and removable covers are provided to give easy access to the main parts of the engine. The transmission gear, which is under the floor, is likewise reached through trapdoors.

The engine is a six-cylinder, compressorless, four-stroke MAN Diesel engine of 70 E.H.P. at

Fig. 113. Diesel motor coach, Halberstadt—Blankenburg Railway. Underframe.

1000 revs./min and 90 E.H.P. at 1250 revs./min. The cylinder bore is 115 mm and the stroke 180 mm. The weight of the engine is about 10 kg per E.H.P. Fuel consumption is 190 to 220 grammes per E.H.P. per hour according to the speed and load. The transmission is of the SLM system ("Schweizerische Lokomotiv und Maschinenfabrik, Winterthur"). Engine and transmission are connected by flexible couplings. The gear box is of light-metal, the gear wheels, shafts, etc. of high grade steel. All shafts rotate in roller bearings. There are four speeds with ratios of 1 : 5.4, 1 : 3.2, 1 : 1.64 and 1 : 1.

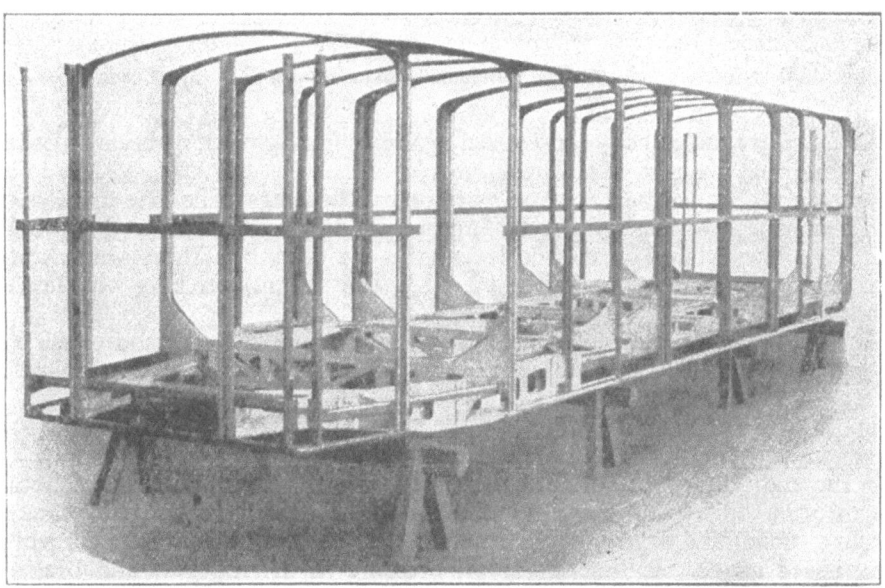

Fig. 114. Diesel motor coach, Halberstadt—Blankenburg Railway. Coach body frame.

The oil pressure of 4—6 atm. required for the clutches is supplied by two separately driven oil pumps, one driven direct by the engine and the other by a shaft of the driving gear. This enables the engine to be started also by the running coach. In addition to the gear wheel transmission there is a reversing gear working at a ratio of 1 : 1.5 and mechanically operated from both driver's cabins. The reversing gear is fitted with a brake operated by a servo-motor by means of oil under pressure; this brake acts as a dead man's brake and is set in working as an emergency brake as soon as the driver leaves his post, the pressure oil being withdrawn from the clutches in the transmission gear, thus breaking the connection between the engine and the driving gear; the engine, then running light, drives one of the oil pumps (even when the coach is at a standstill) and supplies the oil necessary to keep on the brake.

Both axles are driven from the reversing gear by means of cardan shafts.

Fig. 115. Diesel motor coach, Halberstadt—Blankenburg Railway. Coach body.

A large safety factor has been chosen against slipping of the wheels in order to prevent abnormal wear of the wheel tyres.

The axle gear cases bear on the axles in ball bearings. The axle drive consists of bevel gear wheels with a ratio of 1 : 3, corresponding to a max. speed of 40 km/h (25 m.p.h.). These bevel gear wheels can be replaced by others with a ratio of 1 : 2 so as to give a max. speed of 62 km/h (38 m.p.h.) on lines with less steep gradients.

The cooling water is recooled in coolers mounted on the roof and consisting of two elements. By means of a threeway cock the heated cooling water can be used in winter for heating the coach.

Fig. 116 is a plan of the car.

The current for lighting is supplied by a battery of 160 Ah.

On the end walls of the driver's cabins the following apparatuses are mounted: engine throttle, oil switch, reversing handle, hand-brake, oil pressure brake (as dead-man's brake) and sand box; thermometer for cooling water, engine revolution counter, speed gauge, manometer for oil pressure in driving gear and brake, and finally a voltmeter for the battery.

The coach was built by the Herdinger coach factory.

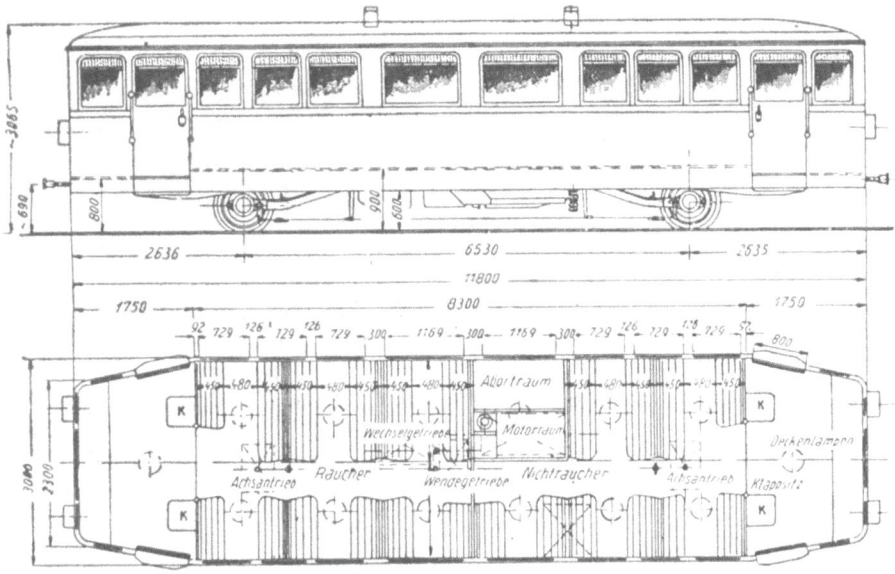

Fig. 116. Diesel motor coach, Halberstadt—Blankenburg Railway (Germany).
Abortraum = Lavatory. — Achsantrieb = Axle drive. — Deckenlampen = Ceiling
lamps. — Klappsitz = Folding seat. — Motorraum = Engine room. — Nicht raucher =
Non-smoker. — Raucher = Smoking compartment. — Wechselgetriebe = Reduction
gear. — Wendegetriebe = Reverse gear.

WUMAG MOTOR COACHES.

Four-wheeled motor coaches.

These motor coaches are built by the "Waggon- und Maschinenbau A.G."
Görlitz (WUMAG).

In order to resist the strong dynamic and atmospheric influences and to afford
a high degree of efficiency, whilst at the same time minimising the weight of the
vehicle, the coach body is made with an iron skeleton.

There is a platform with driver's cabin at each end. A coach of this type is
shown in fig. 117.

In order to prevent the vibrations of the engine being transmitted to the body,
the engine frame is suspended entirely free from the coach body at three points
and the torque of the engine is transmitted direct to the axle (see fig. 118); thus the
coach frame can be of light construction.

The two wheel sets, one of which is the driving set and the other a trailer,
both have radial axles rotating in roller bearings. The engine frame and driving axle
form together one single-axled bogie, thus ensuring smooth rounding of curves.
The whole engine can easily be lifted out of the frame and removed together with
the driving axle, and by putting in a spare engine frame complete with driving-axle
the motor coach is very soon ready for service again.

The engine and machinery is also easily accessible from underneath the coach.

The engine is a six-cylinder unit of 90 H.P. at 1000 revs./min, started electrically and also, in case of need, by hand. Directly coupled to the engine is a dynamo
of 225 watts for supplying the lighting power in combination with an accumulator

Fig. 117. Wumag motor coach.

battery of 160 Ah. The engine cooling water is recooled by a cooler on the roof or may be used for heating the coach.

The transmission from engine to driving axle is mechanical, via gear wheels with five speed variations. The reversing gear is on the driving axle and consists of two bevel gear wheels on the axle and a third on the driving shaft engaging one of the first two according to the direction of travel. The gear wheel transmission is on the *Soden* coupling principle with the ratios 1 : 7.11, 1 : 4.37, 1 : 2.66; 1 : 1.63 and 1 : 1.

The gear wheels are on three shafts running in the gear box on ball or roller bearings and are continuously in mesh. By sliding the gear wheels a little along the shafts they can be engaged with the respective shafts by means of claw couplings without interfering with the mesh.

The control is effected from the driver's cabin by

Fig. 118. Powerassembly of Wumag motor coach.

Fig. 119. Wumag motor coach, German State Railway Company.

electric-pneumatic means, which enables any desired number of motor-coaches to be coupled together and worked from one position. Moreover, one or more engines can be switched in or out while the train is running, thus saving fuel.

The gas handle is provided with a dead-man's grip which when released works the air brake and short-circuits all the engine ignitions. The compressed air is delivered by a compressor driven by the engine and automatically cut out as soon as a certain pressure is reached. This compressed air is used for the air brake, for operating the couplings in the gear box, for the sand box and for the siren.

The fuel tank (of 210 litres) is mounted on the roof, from which the fuel gravitates to the engine.

The main dimensions of this motor coach are:

Tare . . . 16 tons
Engine make: Büssing A.G., Braunschweig.
Engine bore:
  125 mm ($4^{15}/_{16}$")
Engine stroke:
  160 mm ($6^{5}/_{16}$")
Engine power :
  90 H.P.

Engine fuel consumption
　　　　　about 250 g per H.P.h
Seating capacity . . . . . 68 persons
Standing room . . . . . . 30　　,,
Dead weight per seat . . . 238　kg
　,,　　,,　　,,　,, & stand-
　　ing room . . . . 163　,,
Max. speed at 1000 revs./min
　　　　　　72 km/h (45 m.p.h.)

Trial runs have been made with this coach in Germany, Sweden, Norway, Denmark, Holland and Switzerland.

Four-axled WUMAG motor coach of the German State Railway.

WUMAG have built several four-axled motor coaches for the German State Railway (see fig. 119).

The coach body is made of steel throughout. At each end is a driver's cabin which can be reached through separate doors from outside and also from the inside of the coach.

The coach is heated by the cooling water coming from the enignes, the water from each engine warming one side of the coach, so that both compartments are continuously heated; the radiators are under the seats.

The current for the electric lighting is generated during the run by a *Bosch* dynamo, which also charges an accumulator battery for supplying current when the coach is at a standstill.

The bogies are constructed on the system of the "Görlitzer Drehgestell", a four-wheeled bogie (see figs. 120 & 121). The cradle rests on long laminated springs placed longitudinally with 20 mm lateral clearance. Further there are laminated springs above the axle-boxes, with helical springs at each end. The axles rotate in roller bearings. At the ends of the bogies facing the ends of the coaches there are rail guards, and on either side of each driving wheel a sand box. Each bogie has a driving axle and a trailing axle, the former being the end axles of the coach. The engine with accessories is suspend-

Fig. 120. Driving bogie of Wumag motor coach.

Fig. 121. Driving bogie of Wumag motor coach.

ed in a separate frame in each bogie and can easily be removed in its entirety. In the floor of the coach large trap-doors are made to give easy access to the engines and driving gear (see fig. 122). Each bogie carries a fuel tank of 150 litres capacity placed as high as possible, so that the fuel may gravitate to the engines.

The driving gear consists of a gear wheel transmission with five speed variations, a dry multiple-disc coupling, a Cardan shaft with *Kirchbach* linked joint, a safety coupling and a reversing gear mounted on the driving axle. The engine may be started electrically or by hand. The transmission is the same as that described above for the two-axled WU-MAG motor coach. The various speeds are switched in and out by electric or electric-pneumatic means. The multiple-disc couplings and the throw-out of the speed gear are worked by compressed air, the supply of which is regulated electrically.

The coolers for re-cooling the engine-cooling water are mounted on the roof, from which the water is circulated by means of centrifugal pumps.

Fig. 122. Wumag motor coach. Opening in floor.

A motor coach can be run with several trailing coaches and a number of motor coaches can also be coupled together and operated from one driver's cabin.

An illustration of the operating board with electro-pneumatic switches is given in fig. 123.

Fig. 123. Wumag motor coach. Driver's cabin.

The main dimensions of this vehicle are:
Length over buffers. . . . . 21 metres (68' 10³/₄")
Ext. length of car body . . . . 19.7 metres (64'7⁵/₈")

Wheel base of bogies . . . . 3.9 metres (12′ 9½″)
Distance centre to centre of bogies . . . . 13 „ (42′ 7³/₄″)
Seating capacity . . . . 76 passengers.
Standing room . . . . 74 „

Explosion engines
- make . . . . Büssing
- number 2
- cylinders each 6
- bore . . . . 125 mm (4¹⁵/₁₆″)
- stroke . . . . 160 mm (6⁵/₁₆″)
- revs./min, normal 1000
- rating per engine . . . . 90 H.P.
- weight „ „ . . . . 650 kg (1430 lbs.)

Tare . . . . 40 tons.

The following data were obtained from trial runs made on the German State Railways:

Speed	Transmission ratio	Speed		Tractive force at draw hook	
		km/h	m.p.h.	kg	lbs.
3	1 : 2.66	10—20	6.2—12.4	1500—1300	3300—2850
4	1 : 1.63	20—36	12.4—22.3	850—600	1900—1300
5	1 : 1	36—55	22.3—34	400—200	900—450

The average resistance of the motor coach itself is:

speed in km/h	10	20	30	40	50	60	70
resistance in kg/ton	2.7	3.5	4.3	5.5	6.9	8.5	10.2

MOTOR COACHES OF THE HUNGARIAN STATE RAILWAYS *).

The Hungarian State Railways have several motor coaches running, particularly on local lines. These local services used to be maintained with mixed trains of low speeds, but in consequence of competition from motor traffic on the road motor coaches of higher speed were put into service.

Among the first to be taken into use, in 1925, was a four-axled Diesel motor coach built by the "Eisenbahnverkehrsmittel A.G." of Berlin, with a Maybach engine. In 1926 a two-axled petrol motor coach built by Ganz & Co, Budapest, was put into service with a two-axled trailing coach. The satisfactory results obtained with these vehicles led to further orders being placed with the aforementioned and other firms. The "Ungarische Waggon- und Maschinenfabrik in Györ (Raab)" supplied two rail-autobuses for light traffic lines. At the end of 1928 the Hungarian State Railways had 40 motor coaches running, six of which had four axles and the rest two. These two-axled vehicles weigh about 18.4 tons and can seat 46 passengers (15 second and 31 third class). The engine has a rating of 90 H.P. Transmission is pneumatic, on the *Ganz* system, with 4 speeds of 13.1, 23.9, 33.7 and 48.2 km/h, whilst the maximum speed is 55 km/h. The wheel axles are mounted in SKF roller bearings. Buffers and draw gear are of the normal design, and the coach is heated by the returning cooling water.

*) Organ für die Fortschritte des Eisenbahnwesens, 1929, Heft 18/19. 20 Sept. 1929, p. 330.

Two-axled trailing coaches of light-build are used with these motor coaches. Their tare is 12.4 tons and they can seat 56 passengers (15 second and 41 third class). The axles of these trailing coaches run also in roller bearings.

The motor coaches can haul three, two and one trailer up gradients of 1 : 1000, 5 : 1000 and 10 : 1000 respectively. There is a driver's cabin at each end. The coach is lighted by electricity supplied by a 500 watt *Bosch* dynamo, which also charges a 24 volt accumulator battery of 160 Ah and a 24 volt accu battery of 80 Ah in the trailer coach. At present the trailer coach is still heated by a stove, but other heating arrangements are being tried out.

In the motor coaches, in addition to the passenger compartments, there is a luggage compartment with an area of 1480 × 1342 mm; a two-axled goods van of 8.5 tons can also be taken along when the luggage room is insufficient.

Each motor train has a driver and a guard; in case of need the latter can also stop the train.

The rail-autobuses, of which there are two in service, weigh 7.7 tons and can seat 29 passengers each.

Transmission is mechanical, with 4 speed variations. Maximum speed is 60 km/h. The coach is heated by the exhaust gas of the engine. These rail-autobuses have to be turned at the termini as they have only one driver's cabin.

RAIL-AUTOBUS OF THE DANUBE-SAVE-ADRIA RLY.[1]

Fig. 124. Railautobus, Danube—Save—Adria Railway (Hungary).

These rail-auto-buses have been put in service on the lines along the Balaton Lake in order to avoid frequent stopping of the express trains. Passengers are picked up at intermediate stations and carried to a station where the express stops, and vice versa. At the same time a good service is maintained between the numerous health resorts on the Balaton Lake.

Illustrations of these vehicles are given in figs. 124 and 125. They are built just like an ordinary motor omnibus but are provided with wheels for running on rails. There are no buffers or draw gear.

The main dimensions are the following:

Total length . . . . 9.50 m (31′ 2″)
  „ height above rail . . . . 2.80 „ (9′ 2¼″)
Wheel base . . . . 5.70 „ (18′ 8⅜″)
Wheel diameter . . . . 750 mm (2′ 5½″)
Tare . . . . 6.60 tons
Weight on front axle with full load . . . . 4.4 „
  „ on rear axle with full load . . . . 4.5 „

[1]) Organ für die Fortschritte des Eisenbahnwesens 30-1-26. Schienenautobus von Ing. Alexander Pogány, Budapest.

Engine rating . . . . 50 H.P.
Speed on level track
        75 km/h (47 m.p.h.)
Fuel cons. per 100 km 25 litres
Capacity petrol tank 200 „
Seating capacity. . .   29
Weight per seat 227 kg (500 lbs.)

The service schedule is arranged so that the bus can travel at 65 km/h and the express train at 80 km/h. Thanks to the quick starting the bus covers the 85 km between Balaton szt. György and Lepsény in almost 2 hours with 17 stopping places, whilst the express train takes 1 hour 35 min. A speed of 50 km/h can be reached with the bus within 560 metres.

The axles rotate in SKF roller bearings.

The driver's compartment is separated from the passenger compartment and can take, in addition to the driver, also a conductor and if necessary an inspector.

The engine, a four-cylinder, four-stroke, Ganz petrol engine, is mounted in the front of the frame (see fig. 126). The cylinder bore is 130 mm, stroke 160 mm and number of revs./min 1100. The power figures and petrol consumption are shown graphically in fig. 127. The engine has a *Bosch* igniter, *Pallas* carburettor and a *Bosch* electric starting and lighting equipment, whilst further the usual appurtenances are provided. The engine can be started both electrically and by hand.

The cooling water radiator is cooled by a ventilator.

Transmission from engine to driving axle is by means of gear wheels made of chrome-nickel steel. There are five transmission

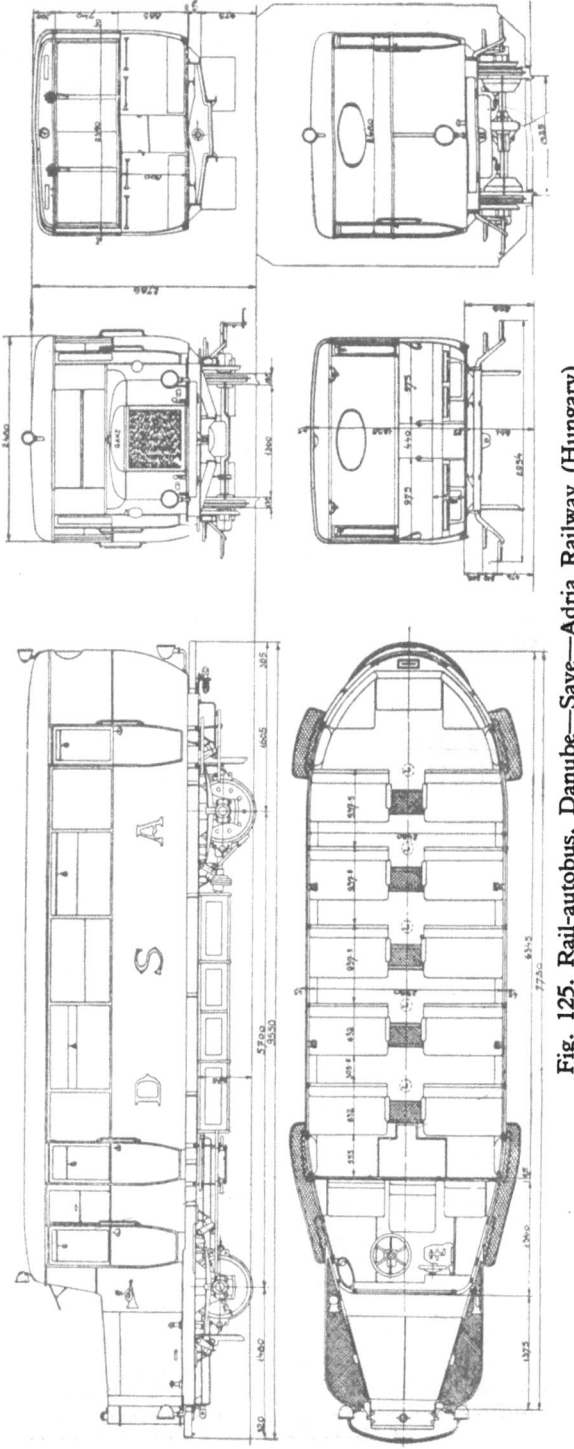

Fig. 125. Rail-autobus, Danube—Save—Adria Railway (Hungary)

gears for the forward direction and two for reversing, giving speeds of 6.2, 11.3, 21.0, 41.1 and 76.3 km/h forward and 11.5 and 21.0 km/h backward.

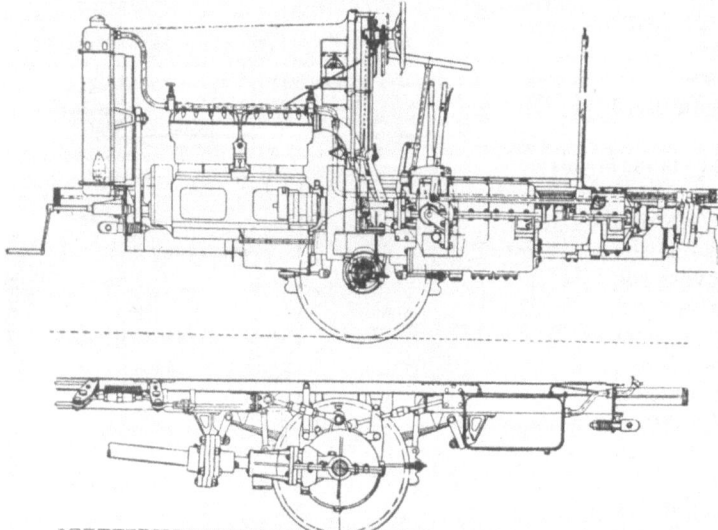

Motion is transmitted to the bevel gear wheels of the rear axle via a cardan joint on the *Rudge-Hardy* system.

The electric light current is supplied by a *Bosch* dynamo connected up parallel with an accumulator battery and driven by the petrol engine. The vehicle is heated by the cooling water, which reaches a temperature of 60—70° C.

A *Bosch* claxon is provided, and as reserve an ordinary motor horn. Further there

Fig. 126. Rail-autobus, Danube—Save—Adria Railway. Power assembly.

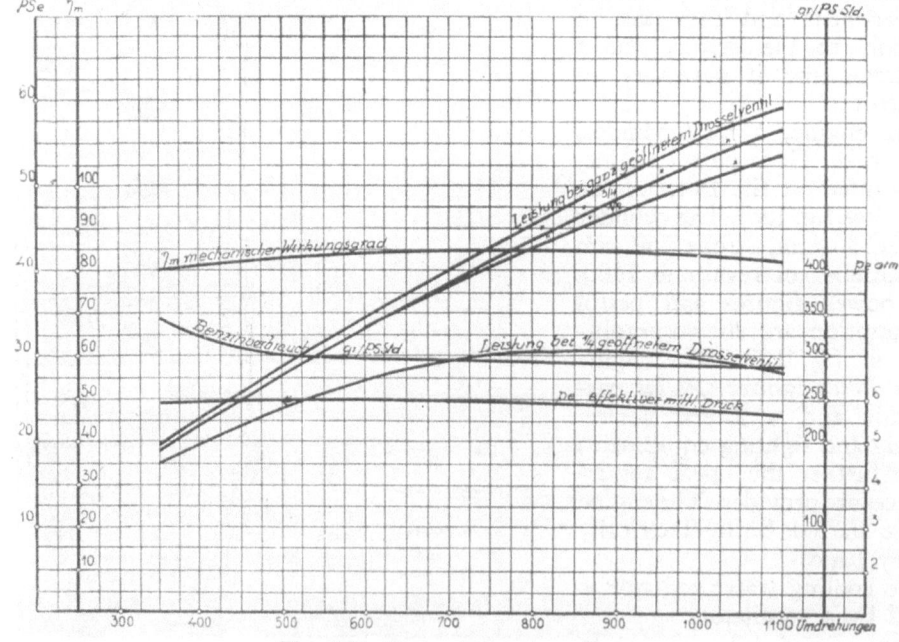

Fig. 127. Data of petrol engine.
Benzinverbrauch gr/PS.Std = Petrol consumption in grammes per horse power hour. — effektiver mittl. Druck = Mean effective pressure. — Leistung bei ganz geöffnetem Drosselventil = Output with fully opened throttle valve. —Mechanischer Wirkungsgrad = Mechanical efficiency. — PSe = Effective horse power. — Umdrehungen = Revolutions.

is a sand box. Luggage can be carried in a box underneath the frame.

The wheels are made of manganese steel on the *Schaffer* system. This material, which wears well, has a tensile strength of 80 kg/mm², with an elongation of 30% and when blue-hot (occurring while braking) does not become brittle. The resistance against wear is five times that of ordinary rolled wheel tyres.

MOTOR COACHES OF THE LILLAFUREDER FOREST STATE RAILWAY (HUNGARY).

These are four-axled narrow gauge coaches built by Ganz & Co of Budapest for 760 mm gauge track. The machinery is much about the same as that of the Hungarian State Railway motor coaches, except that both axles in a bogie are driven via a cardan shaft. An illustration is given in fig. 128.

On the lines over which these vehicles run there are gradients of 38 : 1000 and curves with a radius of 50 m. The max. speed is 56 km/h.

MOTOR COACHES OF THE NETHERLANDS RAILWAYS.

As early as 1923 the Netherlands Railways put three four-axled motor coaches into service. These were built by the "A.E.G." of Berlin in conjunction with "Linke-Hofmann-Werke A.G." of Cologne-Ehrenfeld for the coaches themselves and with the "Nationale Automobil-Gesellschaft" of Berlin for the

Fig. 128. Motor coach train. Lillafüreder Staatliche Forstbahnen (Hungary).

Fig. 129. Motor coach, Netherlands Railways.

engines and driving gear with accessories. They are mainly built of steel and have bogies with one driving axle and one trailing axle each. The wheel axles rotate in journal boxes with disc rings on a system of the "Nordisk Kullager Aktiebolaget". Further the coaches are equipped with a Westinghouse quick-acting brake, an emergency brake, a hand brake, sand boxes operated by compressed air, air-whistle and air-bell at each end of the car.

The coach has two six-cylinder petrol engines of 75 H.P. each. Motion is transmitted to the driving axles via intermediate shafts, changing and reversing gear wheels and cardan shafts with bevel gear wheels. The reversing gear wheels and the friction couplings with which the changing gears are engaged and disengaged are worked by compressed air. The engines and gear-boxes are carried in the frame, the former in a closed compartment on each platform and the latter under the floor, accessible through trapdoors. In each end wall there is a cooler with ventilator, and above each engine is a ventilator shaft with a cowl on the roof.

The engines can be started both electrically and by hand and have electric ignition, for which purpose two accumulator batteries are suspended in boxes underneath the frame; these batteries also supply the current for lighting and for the necessary signal lamps. Each engine is connected up to a *Bosch* dynamo for charging the battery.

Each engine also drives an air compressor for the Westinghouse brake and other appurtenances. When the max. pressure of 5½ atm. is reached in the main air vessel the compressor is short-circuited by a governor.

The petrol for each engine is carried in a tank of 150 litres capacity under the platform at each end. These tanks are each fitted with a small rotary pump for filling, with connection for the suction line on each sole bar of the frame and further a gauge glass and an air-vent pipe passing underneath the coach.

The driver's compartment is on the right-hand side of the engine and each dashboard is fitted with: starting button, reversing crank, speed gear crank, gas handle, Westinghouse brake handle, manometer, whistle and sand-box buttons and a handwheel for the pneumatic bell. A Westinghouse manometer is affixed to the end wall and the wheel of the hand brake is beside the dashboard. The electric switches are mounted on a side wall.

In order that two may be coupled together, these motor coaches have couplings at each end for joining up the main reservoirs and also a connection for connecting up a series of the control conduits.

The coaches are heated by the cooling water from the engines, which flows through pipes along the side walls.

In 1929 seven other four-axled petrol motor coaches were put into service, built by "Werkspoor" of Utrecht, the engines and driving gear being supplied by the "Triebwagenbau A.G." of Kiel. These coaches can be run as one-man coaches and are fitted with a dead-man's grip on the driving handle.

The frame is of steel, with rivetted side and roof plates. Westinghouse, emergency and hand brakes are provided, also pneumatic whistle, foot and pneumatic gongs and a windscreen cleaner.

There are two six-cylinder engines of 115 H.P. each, and one axle of each bogie is driven by each engine. Transmission is on the same principle as that of the other coaches described above.

Each engine with its driving gear is carried in a frame suspended by springs at four points of the car frame, to eliminate unpleasant noises, and each unit is operated by means of compressed air, thus enabling two coaches to be coupled together. The coolers are on the roof.

After the fourth speed has been switched in one of the engines can be cut off, to save fuel.

The petrol tank, with filling openings on the roof, has a capacity of 700 litres, and a gauge is placed in the luggage compartment. The carriage is lighted by current of 24 volts supplied by two dynamos driven by the petrol engines, with the aid of an accumulator battery. The cooling water from the engine serves for heating purposes.

The equipment on the dashboard is as follows:
2 short-circuit switches for each engine,
2 starting buttons,
2 × 2 control lamps for oil pressure and charging of the accu batteries,
reversing handle,
speed variation handle,
gas throttle (dead man's grip),
Westinghouse brake valve handle,
pneumatic bell switch,
switch for cutting off an engine and the gas governor on the dashboard not in use.

Fig. 130. Motor coach of the Netherlands Railways.

On the floor there are:

2 pedals, one for the whistle and one for the bell.

Further there are the following recording instruments:

1 control manometer for the air pressure in the gas-regulating cylinder and the air supply at the first speed,

2 thermometers for the cooling water,

1 manometer for the Westinghouse brake,

1 speed gauge,

2 revolution counters,

1 voltmeter,

1 water pressure manometer for the cooling water.

In addition to these eight-wheeled motor coaches described above the Netherlands Railways have had six four-wheeled motor coaches built by Allan & Co. of Rotterdam (coaches) in conjunction with the "Deutsche Getriebe-Gesellschaft" of Hannover (driving gear) and the "Automobilwerke H. Büssing" of Braunschweig (engines).

There is an operating stand at each end, which can be closed in by a roller shutter. These coaches are likewise designed as one-man coaches, with dead-man's grip.

They are built of steel throughout and have the same brakes and accessories as those described above.

The coach is driven by one six-cylinder engine of about 100 H.P. driving one axle. Speeds are changed by means of compressed air, so that two coaches can be coupled together. The engine and driving gear is mounted in the same way as in the eight-wheeled coaches, and the coolers are likewise on the roof. The petrol tanks in the roof have a total capacity of 300 litres; the petrol level is recorded by a gauge glass.

Lighting and heating arrangements are similar to those previously described.

MOTOR TRAMCARS OF THE NETHERLANDS TRAMWAY COMPANY.

The Netherlands Tramway Company have had some petrol motor tramcars built by "Werkspoor" of Amsterdam on the Werkspoor-de Dion Bouton system (see figs. 131—133). With a view to obtaining the lightest possible construction, the side walls are made to serve as girders, whilst in order to give them sufficient rigidity underneath the windows the latter are not made to open, ventilation being provided by sash-windows above.

The effect of engine vibrations in the car itself is avoided by mounting the body on six helical springs on the steel frame carrying the whole machinery. This frame is carried on the wheels and axles by four laminated springs, while the axles rotate in roller bearings. The springs are fixed to the journal boxes and to the frame in such a way as to need no axle box guides.

The steel frame of the car body is covered on the outside with sheet iron and on the inside with three-ply wood.

There are 21 seats and 15 standing places inside and 19 standing places on the platforms. The main dimensions are:

Petrol engine	cylinders	4
	bore	110 mm ($4^5/_{16}$")
	stroke	150 „ ($5^{15}/_{16}$")
	revs./min	1200
	max. power	56 H.P.
	weight	360 kg (792 lbs.)

	light		fully loaded	
Adhesion weight on driving axle . .	3520 kg	(7750 lbs.)	5800 kg	(12750 lbs.)
Weight on trailing axle	3680 ,,	(8100 ,,)	5600 ,,	(12300 ,,)
Total weight	7200 ,,	(15850 ,,)	11400 ,,	(25000 ,,)
Weight of mechanical part			1350 ,,	(3000 ,,)

Weight of car excl. mechanical part 5850 kg (12900 lbs.)

Wheel diameter
690 mm (2′3³/₁₆″)

Gauge 1435 ,, (4′ 8½″)

Wheel base
3500 mm (11′5³/₄″)

Capacity of petrol tank
100 litres

Speeds: forward:
8.4, 16.8, 28.2, 45 km/h.
backward: 8.4 km/h.

All the necessary pedals, handles and switches are provided at both ends of the car. Both the foot-brake and the hand-brake act on all four wheels, the former by means of internal brake-shoes lined with Ferodo. Inside the car is an emergency brake which acts as a hand-brake on the four brake-drums and when the handle is pulled over the magneto is short-circuited, thus stopping the engine.

Speed gears are changed mechanically, the position of the gears being indicated at the driver's stand. The speed change box is of the normal heavy automobile type. There are four speeds forward and one backward. In order to avoid having to turn the car at the termini a reversing device at the side of the car has to be thrown over.

The engine is a four-cycle, four-cylinder De Dion Bouton petrol engine,

Fig. 131. Motor Tramcar Dutch Tramway Company, Chassis with power assembly and transmission.

and the radiator is cooled by air drawn from the roof. The seats are made to turn up and trap-doors are provided to give easy access to all parts of the engine.

For general overhaul the car body can be lifted entirely, exposing the whole of the mechanism.

The petrol tank is in the engine space and is filled from the outside. The petrol gravitates to the carburettor.

The dynamo for electric lighting is driven by the engine and while the car is running charges the accumulator battery.

The lamp on each platform has a separate switch and inside the car there are 5 lamps, a group of 2, and another of 3, each with a separate switch on the platform. The external lighting consists of a searchlight with a large lamp and an eccentric small lamp at each end, two side lamps and a lamp in each transparent direction board, with separate switch near the driver's seat. The switch for the outside lamps is mounted on the Vitrix board close to the driver's hand and has four positions, viz.

1 — all lamps out,
2 — only side lamps burning,
3 — side lamps and large lamp of the searchlight burning,
4 — side lamp and small searchlight burning.

When the driver's seat is turned up the whole operating board is closed.

The driver can open and close the platform doors from his seat, and when the doors close the movable stop is automatically drawn up.

The car can be heated by means of the exhaust gas from the engine.

A trailer can be hauled of 3 tons tare and 6 tons loading capacity. The wheel base of the trailer is 1.70 metres and the length over buffers 5 metres; the journal boxes are fitted with SKF roller bearings.

The following data are given for the service period of one of these cars from 1st November 1926 to 1st November 1928 in order to give an idea of their efficiency and running expenses.

From 1st November 1926 to 11th October 1927 the car covered 33,566 km.

Fig. 132. Motor Tramcar, Dutch Tramway Company.

From 11th to 19th October 1927, = 8 days, the car was out of service for returning the tyres, cleaning paintwork and revarnishing inside.

From 1st November 1926 to 6th February 1928 44,627 km had been covered, and from 6th to 10th February 1928, = 4 days, the car was in the shops for the engine to be cleaned and new piston springs put in.

From 1st November 1926 to 7th June 1928 59,162 km had been covered, and

Fig. 133. Motor Tramcar and trailer, Dutch Tramway Company.

from 7th to 16th June 1928, = 9 days, wheel tyres were turned again, paintwork cleaned, and ventilators and platform gates fitted on the platforms.

From 1st November 1926 to 1st November 1928 the total distance covered was 76,866 km, whilst the petrol consumption was 28,228 litres and the total oil consumption 1160 litres.

MOTOR COACHES OF THE NATIONAL BELGIAN RAILWAY COMPANY.

These cars have a Maybach Diesel engine with mechanical transmission and were built by the "Eisenbahn-Verkehrsmittel A.G." of Berlin. They have two bogies, one of which carries the engine.

The main dimensions are:

Length over buffers . . . . 21.04 m (69'$^3/_8$")
Length of body . . . . 19.74 „ (64' 9")
Distance centre to centre of bogies . . . . 13.30 „ (43' 7$^5/_8$")
Wheel base of bogies . . . . 3.50 „ (11' 5$^3/_4$")
Wheel diameter . . . . 1.00 „ (3' 3$^3/_8$")
Weight in working order . . . . 40 tons
   „   on engine bogie . . . . 28 „
   „   of mechanical part . . . . 5.2 „
Max. speed . . . . 65 km/h (40.6 m.p.h.)
Seating capacity 35

The car can be driven from both ends. It has a Knorr compressed air-brake and a hand-brake, both working on all wheels. The air is supplied by a compressor to the air vessels and this air also serves for starting.

The car is heated by the cooling water from the engine.

The current for the electric lighting is generated by a dynamo coupled to the engine and connected parallel with an accumulator battery. There are two lamps outside.

The machinery is mounted on one of the bogies. The driving axles are driven by coupling rods from a loose axle, thus safeguarding the driving mechanism from the running shocks of the road.

There are really two parts in the engine bogie, viz. the engine itself and the driving mechanism, each of which is suspended at three points, the two being connected together by a flexible shaft.

On the engine being started, with compressed air, the driving mechanism is switched in and the car begins to move at once. The starting torque is then about 5 times the normal driving torque.

The Maybach Diesel engine is of the four-cycle, six-cylinder type with compressor and a max. rating of 150 H.P. at 1300 revs./min. The fuel consumption between a wide range of load is 185 grammes per H.P. h. Any heavy oils, except tar oils, can be used in the engine. The lubricating oil consumption is about $\frac{1}{2}$ kg per 100 km.

The six cylinders are fed by a fuel pump via a distributor. Lubrication is under pressure. The carter is ventilated. A large centrifugal pump circulates the cooling water.

The compressor works in three stages and is connected to the engine via a flexible coupling. In the first stage the air is compressed to 5 atm., in the second stage to 40 atm. and in the third to 100 atm. The cooling water of the compressor circulates also through the engine cooling jacket. The pressure in the air vessel is regulated by an automatic valve.

The engine and the driving gear are connected by means of a floating shaft with flexible coupling at each end so that some displacement of these parts is possible.

In the driving gear there are four speeds, conical couplings and reversing gear. The driving shaft is above the driven axle. On the latter there are four gear wheels which can be coupled up to the shaft by means of friction couplings worked by oil under pressure, and these gear wheels engage in four pinions on the driving shaft.

The four speeds are 10, 19, 40 and 61 km/h.

The driven shaft in the carter has a conical pinion which meshes into one of the two bevel gear wheels on a second shaft, according to the direction of travel desired.

At both ends of the loose axle there are cranks at 90° from each other, and these drive the driving wheels via coupling rods.

The cooling water radiator, consisting of several pipe series, is on the roof and the water is sufficiently cooled by the wind set up while the train is running, but there are also four ventilators which can be started as desired.

There are two tanks of 160 litres each with oil-level float. The oil is fed to the engine by a pump.

Compressed air of 100 atm. is stored in three air vessels of 115 litres each, one of which is a reserve.

In each driver's cabin there are handles for operating engine and mechanism.

Gear can only be reversed when the car is at a standstill. The driving handle is a dead-man's handle. The dashboard moreover comprises a speed gauge, a distance thermometer for the cooling water, two manometers (one for starting air and the second for the combustion air) and 2 manometers for the air-brake.

Trials were made with these motor coaches to see how long it takes to reach a speed of 35 km/h on gradients, with the following results:

Train weight	Gradient	Distance from starting	Time from starting
motor coach alone (43 tons)	0 1 : 100 up 1 : 100 down	220 m 430 „ 150 „	35 sec 63 „ 25 „
with one trailer coach (60 tons)	0 1 : 100 up 1 : 100 down	330 m 1450 „ 190 „	50 sec 180 „ 31 „
with two trailers (77 tons)	0 1 : 100 up 1 : 100 down	450 m [1]) 240 „	70 sec [1]) 38 „

[1]) 35 km/h could not be reached on this gradient with two trailing coaches. A speed of 19 km/h was reached in 30 sec over a distance of 100 metres.

One of these motor coaches was further subjected to the following tests:
1. Twelve four-wheeled coaches were coupled on, making the train weigh 200 tons. This train was easily started on the level track, sharp switch curves were taken and a speed of 8 km/h was reached up a gradient of 1 : 100.
2. A load of 388 ton was coupled on, consisting of goods vans and four and six-wheeled cars, totalling 49 axles. This train was easily started up three times and on the level a speed of 10 km/h was reached.

These motor coaches are destined for the Ghent-Eecloo line, on which four return journeys are to be made per day.

MOTOR COACH OF THE PARIS SUBURBAN DISTRICT RAILWAY.

This coach was built by Schneider & Cie of Paris for the district Seine et Oise, where many gradients (as much as 6 : 100) and curves of small radius (20 metres) occur. The track is of normal gauge and the construction gauge is the same as for tramways.

The engine, 60 H.P., is contained in a closed box placed in the middle and at one side of the vehicle and is accessible through a removable panel on the outside (see fig. 134). The coolers are on both sides of the engine and ventilation is obtained by a partial vacuum in the engine box. The engine is of the explosion type with four cylinders, bore 135 mm and stroke 170 mm. Petrol or a similar fuel may be used. The rating is 60 H.P. at 1000 revs./min and 70 H.P. at 1200 revs./min; the normal engine speed while the train is running is 700—1250 revs./min. The engine is equipped with an automatic carburettor, electric ignition and a dynamo for starting and for lighting. The ventilator is on the flywheel.

The coach can be driven from either end.

The speed variation gear is on the *Fieux* system and the driving mechanism is adjusted for 4 speeds, viz. 14, 22, 40 and 30 km/h at 1200 revs./min. Intermediate speeds are obtained by adjusting the engine speed.

Both axles are driven via cardan shafts driving a set of bevel gear wheels on each axle. All axles rotate in ball bearings.

For smooth rounding of curves the axles are made to move sideways, by means of spring suspension and balances.

In addition to the hand-brake there is an air-brake operated by means of compressed air from a compressor driven by the engine.

The coach is fitted with sand boxes for both travelling directions, heating arrangement for exhaust gas or cooling water, and electric lighting from an accumulator battery.

Fig. 134. Schneider Motor coach, Réseau de Grande Banlieue.

The two driver's cabins are closed in and locked with a safety lock. On leaving one cabin the driver has to lock it before he can take the key out to open the other cabin door.

The tare of the vehicle is 10 tons (5 tons on each axle) and the fully loaded weight is 13 tons.

The following are the speeds attainable on various gradients and with a trailing load of 8 or 16 tons.

Gradient	Motor coach alone		With 8 tons load		With 16 tons load	
	km/h	m.p.h.	km/h	m.p.h.	km/h	m.p.h.
1 : ∞	50	31	50	31	50	31
1 : 100	50	31	45	28	42	22.5
2 : 100	38	24	32	20	20	12.5
4 : 100	20	12.5	18	11	12	7.5
6 : 100	18	11	11	7	—	—

The fuel consumption is 1½—2½ litres per 1000 tkm. On the St. Germain-Poissy line it is 45 litres per 100 km, and on the various lines of the North-Western division it averages 37 litres per 100 km.

MOTOR COACHES OF THE FRENCH STATE RAILWAYS.

The same principle as described above has been applied to several other motor-coaches, i.a. to rebuilt cars of the French State Railways.

An old second-class coach (fig. 135) with four compartments was rebuilt by breaking away two of the compartments at one end and fitting that space up for

Fig. 135. Schneider Motor coach, French State Railways.

the machinery. The frame was left unaltered, only one axle being replaced by a stronger driving axle with chain transmission. This second class motor coach is run with a third-class trailer on the Mortagne-St Gauburge line, a distance of 35 km, which is covered in 70 minutes, with 9 stops, at a speed of 35 km/h. The main dimensions are:

Engine rating . . . . 60 H.P.
Cylinders 4
Bore . . . . 135 mm (5⁵/₁₆")
Stroke . . . . 170 „ (6¹¹/₁₆")'
Weight of motor coach empty . . . . 14½ tons
   „   „   „   „   fully loaded . . . . 16 tons
   „   „   trailer empty . . . . 9½   „
   „   „   „ fully loaded 11   „
Fuel consumption . . . . 0.65 litre per km
                                    or 2.4 litres per 100 tkm.

There are four forward speed gears and one backward, the coach having to be turned at the termini. The speeds and tractive forces are tabulated below:

	Speed				Tractive force* at driving wheel periphery	
	at 1000 revs.		at 1200 revs.			
	km/h	m.p.h.	km/h	m.p.h.	kg	lbs.
1st speed	12	7.5	14	8.5	1230	2700
2nd „	20	12.5	21	13	720	1500
3rd „	34	21	40	25	440	970
4th „	50	31	60	37	280	600
reversed	12	7.5	14	·8.5	—	—

*) Mechanical efficiency is taken at 85%.

In view of the satisfactory results obtained with this motor coach several others have been rebuilt by Schneider & Cie for the French State Railways.

RENAULT-SCEMIA [1]) PETROL MOTOR COACHES.

These cars (see figs 136—141) are normally built to run in both directions and thus have driver's cabins at each end. Two engines of different powers can be installed and the mechanical transmission works on one or two axles, thus giving four types of motor coaches for any gauge. The characteristics of each type are given below:

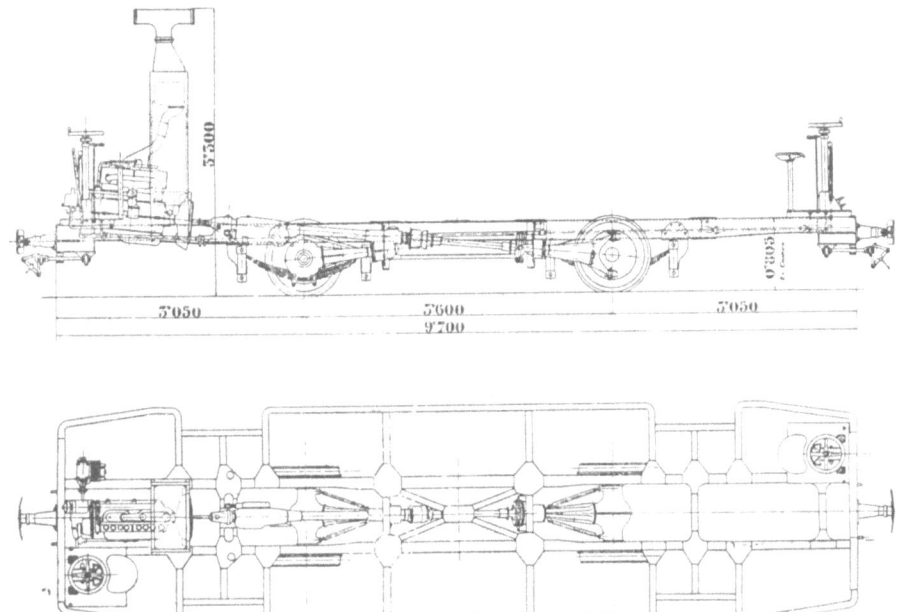

Fig. 136. Renault-Scemia Motor coach. (Coach body removed). — En charge = loaded.

[1]) Société Scemia, Paris.

Type	Driving axles	4 Cylinder engines				Revs. per min	Effective H.P.	Approx. weight in tons for gauge	
		cylinder bore		stroke				1 m	1.44 m
		mm	inch	mm	inch				
RS 1	1	100	$3^{15}/_{16}''$	160	$6^{5}/_{16}''$	1500	45	7.4	7.6
RS 2	2	100	$3^{15}/_{16}''$	160	$6^{5}/_{16}''$	1500	45	7.7	7.9
RS 3	1	125	$4^{15}/_{16}''$	160	$6^{5}/_{16}''$	1250	58	7.5	7.7
RS 4	2	125	$4^{15}/_{16}''$	160	$6^{5}/_{16}''$	1250	58	7.8	8.0

The following table gives the running conditions of the motor coach hauling a load in both directions on various gradients on a straight track.

Type of car	Speed gear No.	Speed		Max. gradient in $^0/_{00}$	Loads hauled in tons on gradients in $^0/_{00}$						
		km/h	m.p.h.		10	20	30	40	50	60	70
RS 1 and RS 2	1	9	5.5	73	58.5	28.5	16.5	10	5.5	—	—
	2	16	10	45	32.5	13.7	6	—	—	—	—
	3	22	13.5	32	20.5	6.5	—	—	—	—	—
	4	40	25	11.5	—	—	—	—	—	—	—
RS 3 and RS 4	1	9	5.5	97	81	45	25.5	17.4	11.4	7.6	5
	2	16	10	60	46	21.5	11.4	6	2.4	—	—
	3	22	13.5	42	30	12	4.8	—	—	—	—
	4	40	25	17	6.5	—	—	—	—	—	—

The air for cooling is drawn from the roof through a shaft with two openings; the one opposite the direction of travel is closed by an automatic shutter.

The multitubular radiator behind the engine is enclosed, as is also the engine, in a casing with an opening leading to a ventilator driven by the flywheel. The air drawn in by this ventilator passes through the nest of tubes and the casing. The cooling water is circulated on the thermo-siphon system and recooled by air via the air shaft mentioned above.

The engine can be started electrically and also by hand by means of a shaft with detachable crank.

The friction clutch consists of a cone lined with Ferodo, which does not exercise any axial pressure on the crankshaft. It allows of a gradual starting of the coach and easy change of speeds.

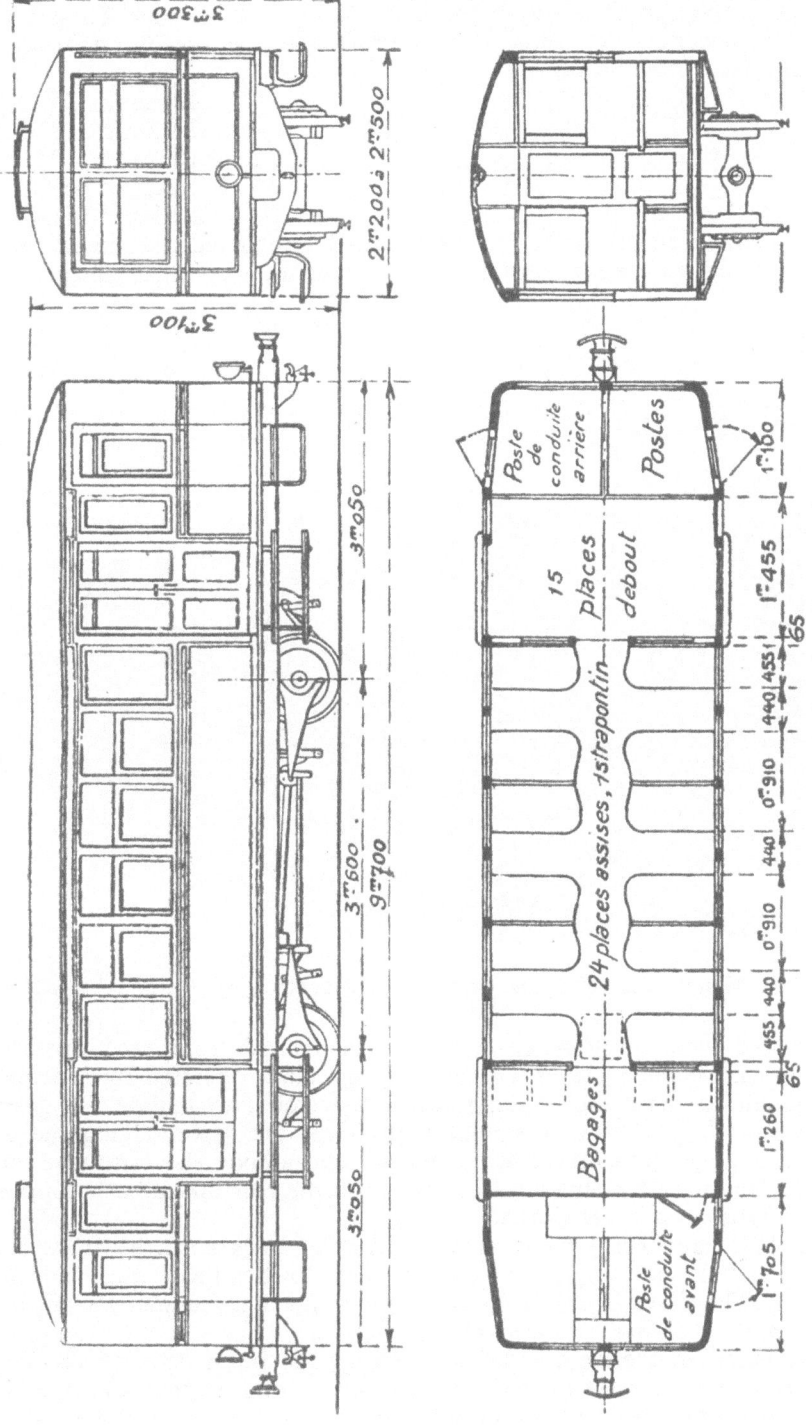

Fig. 137. Renault-Scemia Motor coach.

Bagages = Luggage. — Places assises = Seats. — Places debout = Standing places. — Poste de conduite arrière = Rear driver's cabin. — Poste de conduite avant = Front driver's cabin. — Strapontin = Folding seat.

The gear box has four speeds forwards and one backwards. These speeds are obtained by movement of three sliding gears and the speeds attainable with a normal number of revolutions are:

4th speed, direct coupling . . . . 40 km/h (25 m.p.h.)
3rd   „ , ... 22 km/h (13.5 „)
2nd   „ , ... 16 km/h (10   „)
1st   „ , ... 10 km/h (6   „)
reverse . . . . 10 km/h (6   „)

The fourth speed being the one most usually employed, the direct coupling is an appreciable advantage on account of the elimination of intermediate gears.

Fig. 138. Motor coach, Chemins de fer départementaux du Tarn. (France).

The speed variation gear (patented) is controlled from each driver's cabin by means of an oscillating lever working a single shaft operated by a double rotary movement to select and a longitudinal movement to engage the gears. The hand lever may be locked in the neutral position, the driver taking the key with him when changing over to the other cabin, so that nobody else can move the lever.

The cardan shaft transmission acts direct on the rear driving axle (opposite the engine) by means of a bevel clutch.

In the RS 2 and RS 4 types with two driving axles both the front and rear axles are rotated via a gear wheel on the longitudinal shaft at the first cardan joint. The force is transmitted direct in the longitudinal axis of the frame via a ball joint.

Transmission to the driving axles is via special bevel gear wheels consisting of a bevel pinion meshing with two crown wheels idling on the axle. A dog clutch slidable but not rotatable on the axle can be engaged with either of the crown wheels, the axle being rotated in either direction at the same speed and with the same direction of rotation of the bevel pinion. With this arrangement only one set of gear wheels is in action for each direction of travel. The efficiency of the transmission is the same for both directions and for direct drive is very high, by reason of only one bevel gear being in action for each axle.

In order to ensure proper reversal of motion, a simple disc clutch disconnects the two axles at the moment of reversing. A hand lever in the rear driver's cabin (opposite the engine) operates simultaneously the clutch and the reversing gear; this lever can be detached and taken along to the other cabin when changing round, thus avoiding any interference by others.

The motor coach is fitted with two independent brakes, viz.:

1. a hand-brake acting on all four wheels, and
2. a foot-brake acting on the driving mechanism.

The hand-brake shoes lined with "Ferobestos" work on the inside of drums mounted on the four wheels. The shoes and drums are placed on the outside of the wheels, so as to be easily accessible.

A pneumatic brake (compressed air or vacuum brake) can also be installed, acting on the same brake-shoes as the hand-brake.

The cast iron foot-brake (auxiliary brake) shoes act on the outside of a pulley mounted on the longitudinal cardan shaft; only the driving wheels are affected.

Sand boxes are provided for both directions of travel.

The axles, which are of a special design, rotate in roller bearings.

There is no differential.

In order that the wheels can be taken off for inspection of the axles and bearings they are mounted on tapered split hubs fixed to the axle by tightening a

Fig. 139. Motor coach and trailer, Chemins de fer de la Banlieue de Reims (France).

Fig. 140. Motor coach, Chemins de fer de grande banlieue (France).

locked nut. Locking rings check the longitudinal displacement of the hubs. The wheel tyres are the same as those of ordinary cars.

The vertical suspension at each axle bearing consists of a laminated spring attached to the journal box. The frame is suspended at the extremities of the laminated springs by means of helical springs, which latter have to take up the light vibrations which are not absorbed by the laminated springs. Thus all the advantages are obtained of a double suspension as employed in modern railway and tramway vehicles.

In order to eliminate the transmission of lateral shocks from the axles direct to the frame, which shocks are mainly caused on entering a curve not preceded by a parabolic transition, the links of the vertical suspension springs are elastic, being made of spring steel. The lower ends of these links fit into a metal block through which the axis of oscillation of the link passes. The top part supports the ends of the laminated spring. The links allow of a displacement of the coach body with respect to the axles. Thus in a very simple manner the same results are obtained as are derived from the oscillating suspensions employed in a large number of tramway cars on the double suspension system.

This suspension, having no joints whatever, is free from friction and works quite smoothly.

The usual buffer and draw-gear is fitted at each end of the coach.

The following fittings and appliances are provided in each driver's cabin:

clutch, brake and accelerator pedals,

speed-changing and reversing lever,

hand-brake lever,
two sand-box levers,
handwheels for fuel and air control,
a push-button for the electric starter,
bell and whistle buttons.

In the cabin at the engine end is a speed gauge and in the other cabin a tacho-meter recording the engine speed and also the reversing lever.

The dynamo-starter unit for starting the engine and lighting the carriage has two induction coils and two collectors. It is a combination of two electric motors, a starting motor and a power generator, the latter of the constant voltage type with three brushes. This combination is mounted close to the end of the crank shaft; it

Fig. 141. Motor coach and trailer, Chemins de fer de grande banlieue (France).

is provided with an automatic speed variation allowing it to run 8 times as quick as the engine at the moment of starting and at the same speed as the engine for re-charging the accumulators and supplying the lighting power as long as the engine is running.

The dynamo recharges an accu battery of a capacity of 80 Ah at 12 volts and from the two power sources the following lamps are lighted:
2 lamps of 50 C.P. in the headlights,
2 ,,   ,, 10 ,,   ,, ,, driver's cabins,
2 ,,   ,, 10 ,,   ,, ,, luggage and mail compartments,
3 ,,   ,, 25 ,,   ,, halophane globes, two in the passengers' compartment and
                 one on the platform.

At each end of the motor coach is a contact for lighting the trailer car, via a flexible cable.

The lighting capacity from the dynamo and battery is 150 C.P. with lamps of 1 watt per C.P. In the cabin next to the engine is a switchboard with meters for all

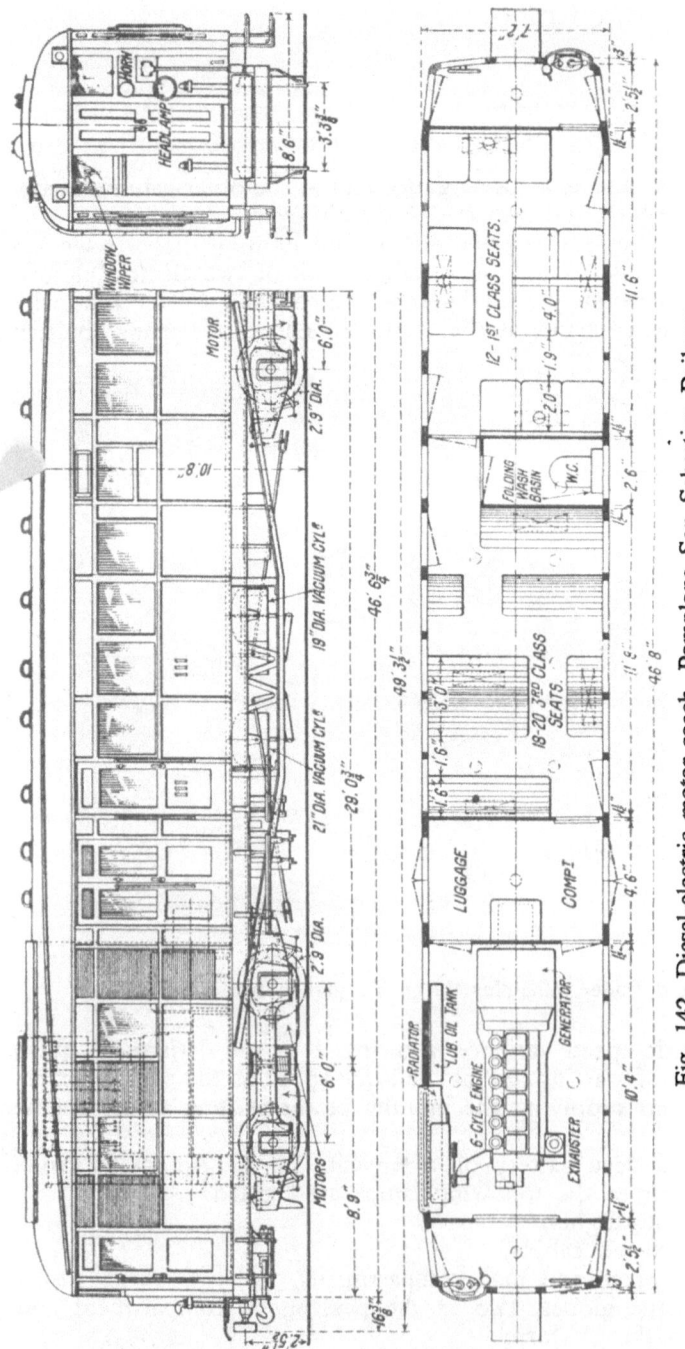

Fig. 142. Diesel-electric motor coach, Pomplona San Sebastian Railway.

internal lighting and for the headlight at that end of the car; the head lights at the other end are served from a small switchboard in the other cabin, on which the engine tachometer and a lighting interrupter are also mounted. Each switchboard has a contact for the lights in the trailer car.

An electric bell is fitted in each cabin, with push-buttons on the entrance platform and in the luggage compartment.

Two kinds of warning signals are provided at each end:
1. an air whistle, and
2. a hand-gong for use in busy traffic.

The motor coach may be equipped with a compressed air brake, semi-direct and automatic, or a vacuum brake acting simultaneously on all four wheels of the motor coach and also on the tyres of the wheels of the trailer car; for this purpose hose couplings are provided at each end of the car. The air compressor and vacuum pump with regulator are placed in front, perpendicular to the engine, by which they are driven direct.

The passenger compartment is heated by the exhaust gas circulating through heaters under each seat; the circulation is regulated by a valve in the exhaust pipe.

The frame of such a motor coach is shown in fig. 136, and fig. 137 is a general plan. Figs 138—141 show some specimens of Renault—Scemia motor coaches.

The petrol consumption of a type RS 4 coach on the Poissy—St. Germain line is said to be about 35 litres per 100 km without trailer.

DIESEL-ELECTRIC MOTOR COACHES OF THE PAMPLONA-SAN SEBASTIAN RLY (SPAIN)[*]).

The Pamplona-San Sebastian railway is a local line with steep and long gradients and reaches in the Pyrenees to an altitude of more than 2000 ft. above sea-level. On this line three four-axled Diesel-electric motor coaches have been put into service, built by Wm Beard more & Co. Ltd of London and Glasgow together with the Abbey Works of Clayton Waggons Ltd of Lincoln, where the coach body was built and the whole assembled. The electric equipment was made by the English Electric Company.

Illustrations of this type of motor coach are given in figs. 142 and 143.

The motor coaches can draw trailer coaches or be coupled in pairs and operated from one position. With these three motor coaches a service is maintained for which formerly seven steam engines were used.

The main dimensions are:
Gauge . . . . . . 3' 3³/₈" (1 metre)
Length over buffers . 49' 3" (15.011 m)
Width . . . . . . 7' 2" (2.184 m)
Seating capacity 1st class 12
Seating capacity 3rd class 20
Engine: cylinders 6
bore . . . . . 6½" (165 mm)
stroke . . . 9" (229 mm)
rating . . . . 200 E.H.P.
revs./min . . 1250

The coach comprises an engine-room, in which the engine and electric generator are placed, a luggage compartment, one 1st class and one 3rd class compartment and a lavatory. At each end is a driver's cabin and a gangway. The cars have a vacuum brake, vacuum sander, vacuum horn and electric lighting.

The engine is a Beardmore, four-stroke, compressorless Diesel engine. The speed

Fig. 143. Diesel electric motor coach, Pamplona San Sebastian Railway.

*) The Railway Gazette, Oct. 19, 1928, p. 490.

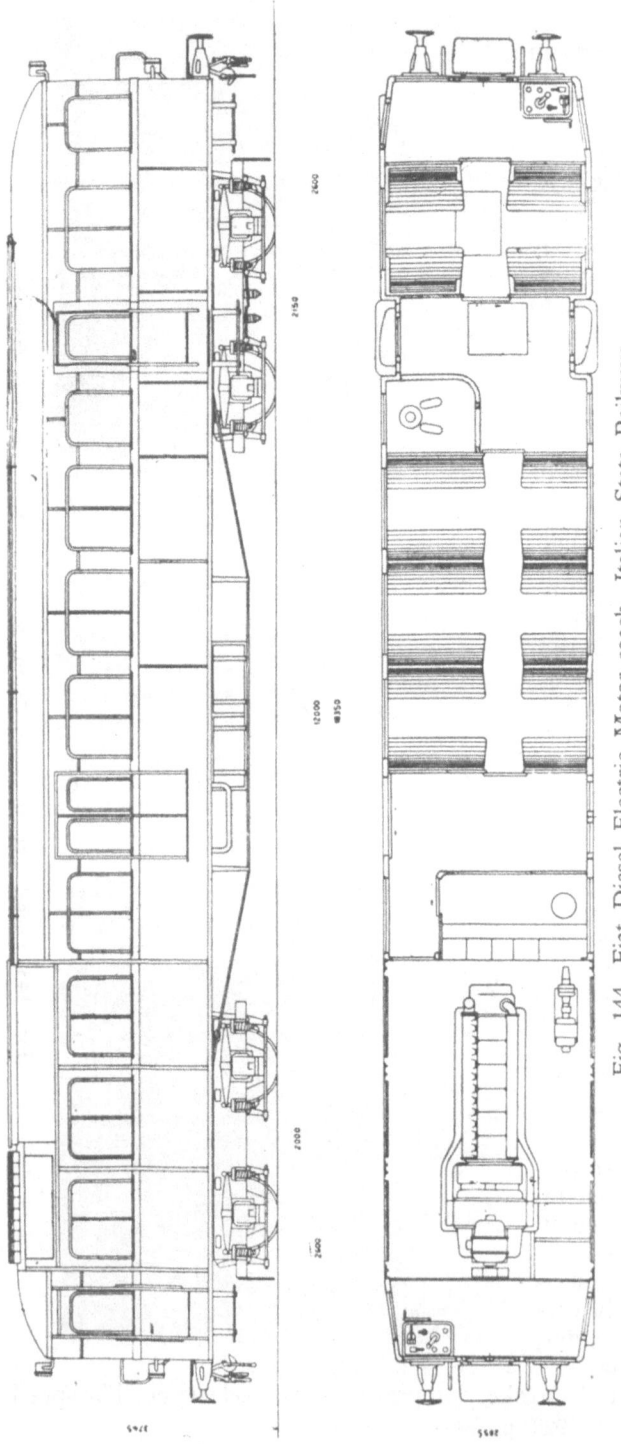

Fig. 144. Fiat Diesel Electric Motor coach, Italian State Railways.

can be regulated between 600 and 1250 revs./min, with a maximum of 1400 revs./min.

The cooling water circulates through the engine cylinder jackets and then through the cylinder heads. The cooler is of the *Reliance* patent tubular type and is placed in one of the side walls of the coach, with ventilators. For the cooling of the lubricating oil a special cooler is provided.

When a defect occurs in the cooling water or lub. oil circulation a signal lamp burns in each of the driver's cabins.

The engine is started electrically by running the generator as a motor with current from an accumulator battery, which also serves for the lighting and supplies the operating power.

The generator is coupled direct to the Diesel engine and produces 135 kW at 1250 revs./min. The accu battery charged by the generator has a capacity of 140 Ah with a discharge time of 5 hours.

The four axles are driven by four motors, two in each bogie. Each motor has a capacity of 50 H.P. (hourly rating) at 500 volts tension.

The ratio of the gear wheel transmission from the motorshafts to the driving axles is 70 : 14.

FIAT DIESEL-ELECTRIC MOTOR COACHES.

These motor coaches were built by the "FIAT Stabi-

limento Grandi Motori" of To-
rino for the Italian State Railways for service on lines with
light traffic and for local lines
with normal and narrow gauge,
in order to get the advantages
of electric traction without the
expense of building sub-power-
stations and laying electric wires.

One of these motor coaches,
the Fiat TA 180, is shown in
figs. 144 and 145, and its main
dimensions are:

Gauge . . . . 1445 mm (4' 9")
Length of body 17.20 m (56' 5")
Length over buffers
        18.15 m (59' 6½")
Width of body 2.86 m (9' 4⅝")
Max. load on driving axle
          12.6 tons
  "   "   " trailing axle
          9.4 tons
Weight in working order 44   "
Steepest gradient that can be
       climbed 4 : 100
Max. speed 60 km/h (37.5 m.p.h.)
Smallest radius of curve that
  can be taken 80 m (88 yard)
Seating capacity 1st class . 8
  "   "   3rd " . 30
Standing room for 20

There are two four-wheeled
bogies with double spring suspension. The axles of one bogie
are driven.

A driver's cabin is provided
at each end (fig. 146), and the
engine room is at one end (fig.
147 a, b). The internal arrangement may be seen from the illustrations.

The current for electric lighting is supplied from an accumulator battery. The coach is
heated by means of the cooling
water from the engine.

The engine is a four-stroke,
compressorless, Diesel engine
of the Fiat V 206 type with six

Fig. 145. Fiat Diesel-electric Motor coach, Italian State Railways.

cylinders, specially designed for locomotives and motor vehicles, with the following main dimensions:

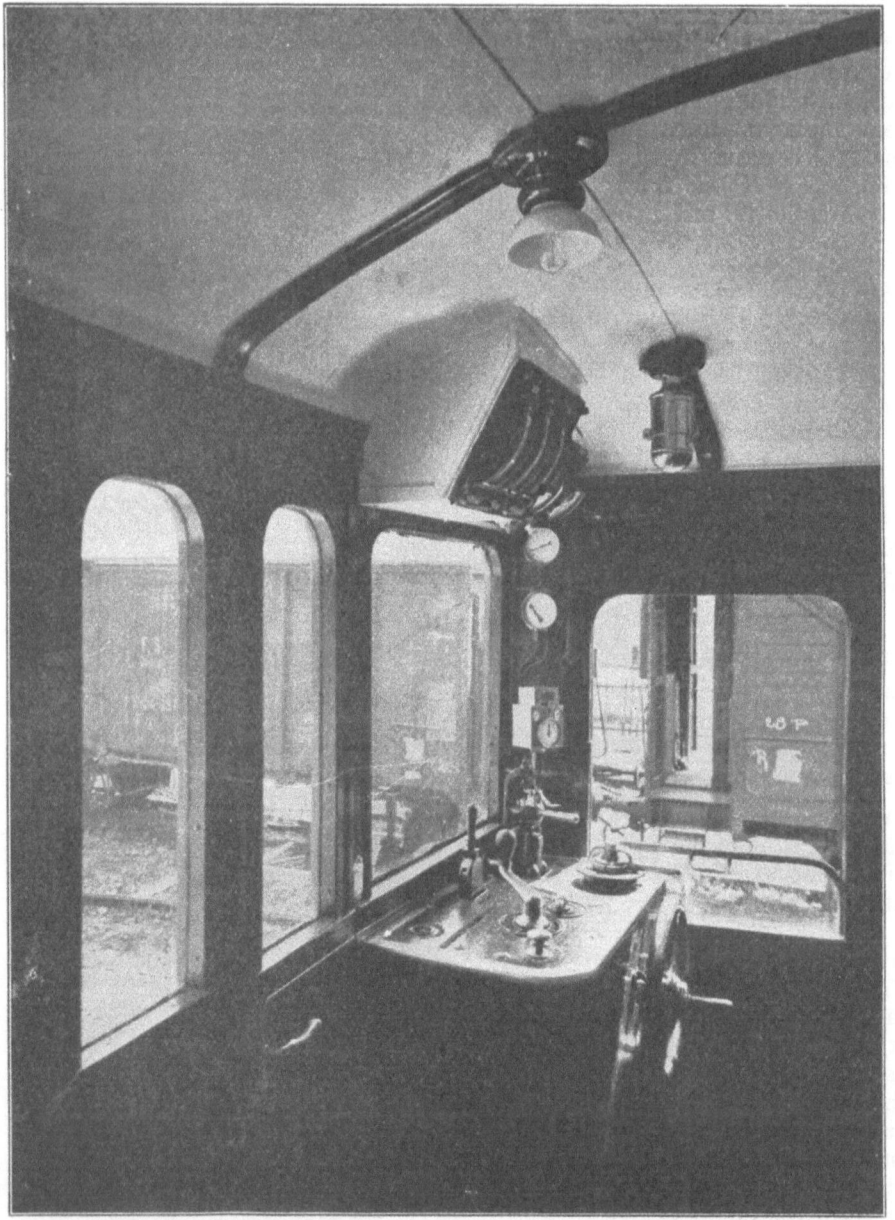

Fig. 146. Diesel-electric motor coach, Italian State Railways. Driver's Cabin.

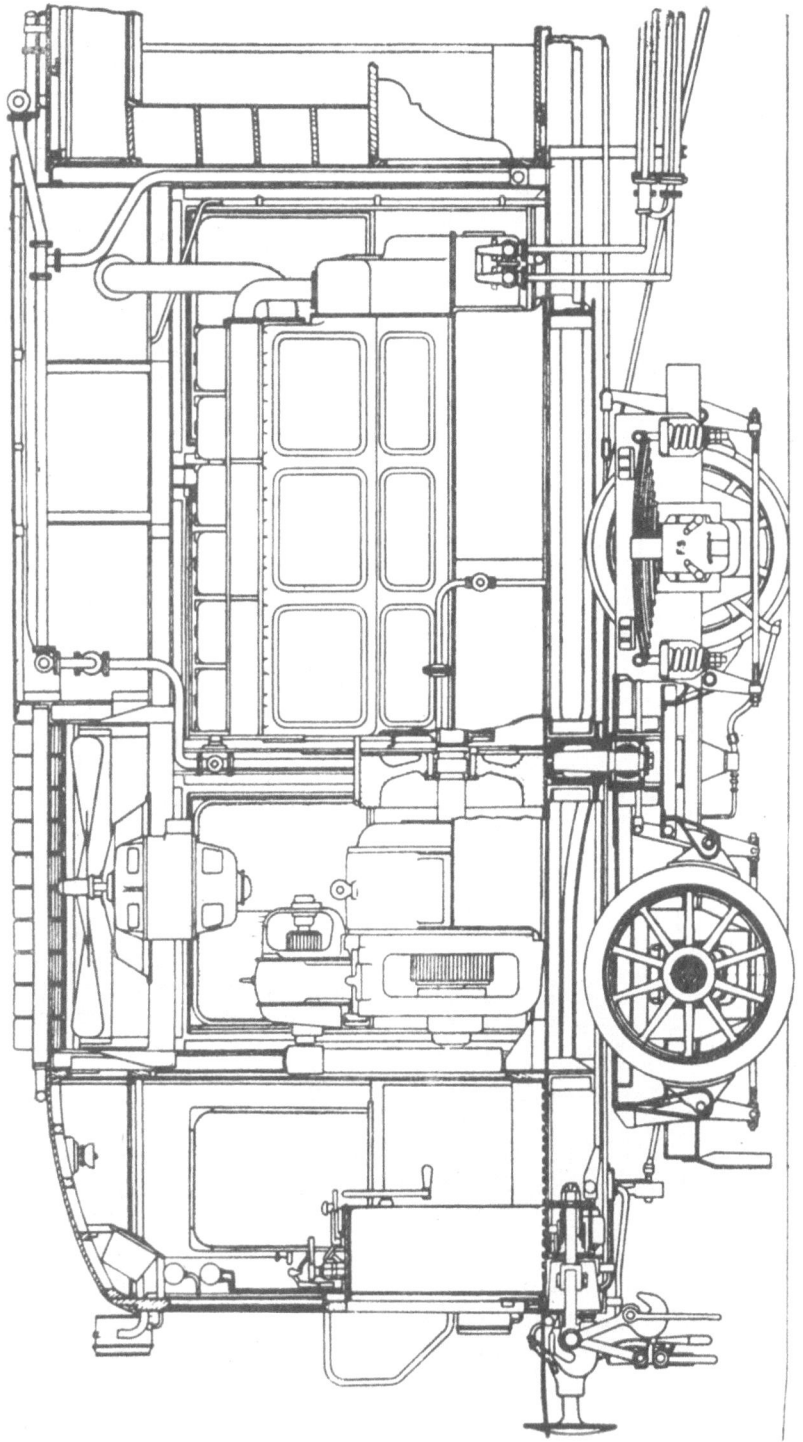

Fig. 147a.
Fiat Diesel-electric Motor coach. Engine room.

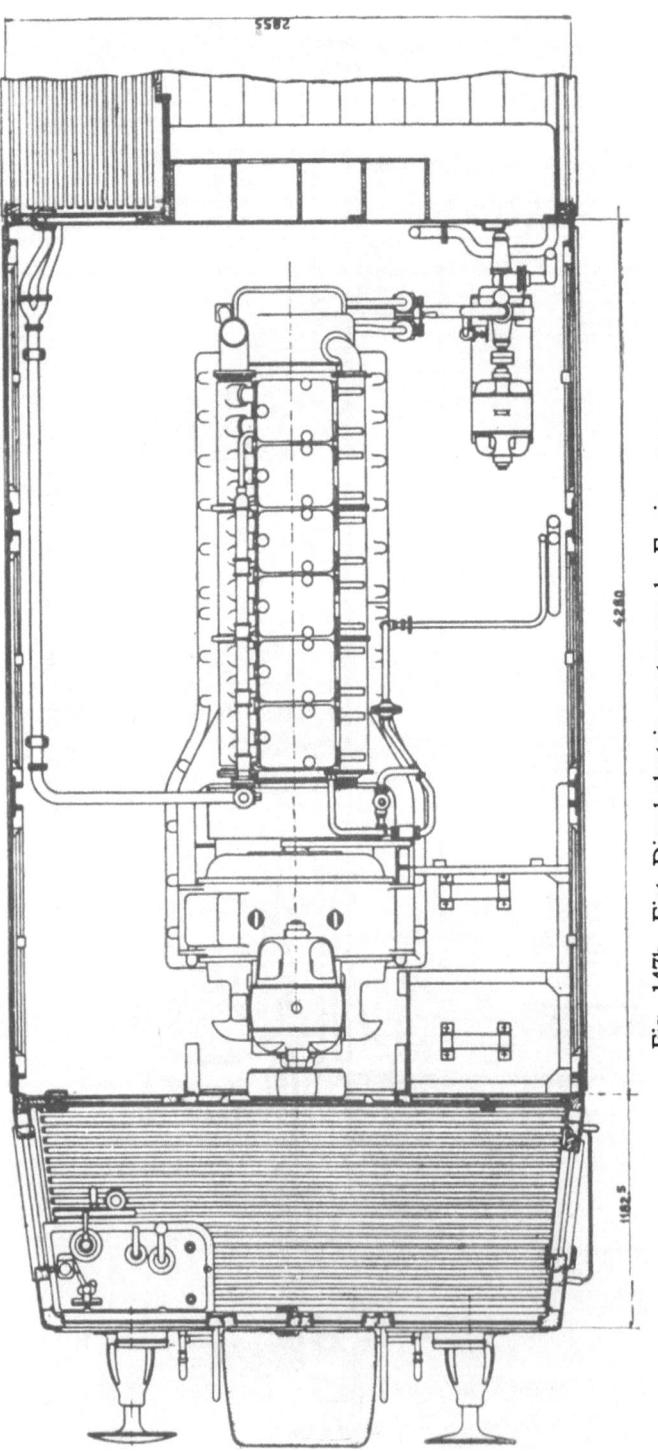

Fig. 147b. Fiat Diesel-electric motor coach. Engine room.

Bore . . 200 mm (7$^7/_8$")
Stroke . 270 ,, (10$^5/_8$")
Normal rating 180 H.P.
Max. ,, 200 ,,
Normal revs./min 750

The engine is mounted together with the dynamo on the frame, which gives great rigidity to the whole. The main dimensions of the generator are:

Output . . . . . 85 kW
Revs./min . . . 750
Max. voltage . . 360
Amperage . . . 300

The generator can be overloaded for an hour up to 100 kW at 450 volts.

There are two driving motors, both in one bogie, each driving one axle.

An accumulator battery supplies the current for starting the Diesel engine, for electric appurtenances and for lighting the coach.

The following accessories are installed:

a) Electrically-driven pump for cooling water; capacity 250 litres per min; power required 2.2 kW; revs. 2400 per min.

b) Electric ventilator for cooler; capacity 700 m³ per min; air pressure 10 mm water gauge; power required 4.5 kW; revs. 370 per min.

c) Compressor for air-brake; max. power 335 litres per min; max. pressure 7 atm.; power required 2.6 kW.

d) Cooler with 68 m² cooling surface.

e) Tubular cooler with natural ventilation, mounted on roof of coach; cooling surface 26 m²; in winter this can be disconnected.

f) Heating arrangement comprising a three-way cock for conducting the cooling
water through pipes in the carriage.

g) Fuel tank of 300 kg capacity, sufficient for 8 hours under full load.

h) Silencer for the exhaust gas; capacity about 400 litres.

DIESEL-ELECTRIC MOTOR COACH OF THE SWISS FEDERAL RAILWAYS.

This motor coach, built by the Société Industrielle Suisse of Neuhausen and
shown in fig. 148, is in regular service on the Brugg-Wohlen line, of normal gauge.

The Diesel engine is a compressorless, four-stroke Sulzer engine of 250 H.P.
It is coupled direct to a DC dynamo of Brown Boveri & Cie. The traction motors
are coupled up in such a way that the travelling speed can be varied within a wide

Fig. 148. Diesel-electric motor coach, Swiss Federal Railways.

range with a constant speed and power of the Diesel engine. When travelling down
gradients the Diesel engine can be idled, as shown in fig. 149. In regular service
the fuel consumption is 1 litre per km with a trailing load of about 20 tons.

There is a driver's cabin at each end, a luggage compartment of an area of about
7 m², and two passenger compartments with a total seating capacity for 50 passengers.

Total weight of motor coach, empty . . . . 57 tons
Adhesion weight . . . . 28.5 tons
Max. speed on own power . . . . 60 km/h (37.5 m.p.h.).
Max. admissible speed as trailer . . . . 75 „ (47 m.p.h.).

The fuel consumption on the test bed was 188 grammes/H.P.h at full load
with gasoil of 10500 kcal/kg.

On the trial track Wallisellen-Romanshorn (represented in fig. 149) the total
fuel consumption for the return trip was 74.8 kg with a train weight of 76 tons
over a distance of 147.2 km; i.e. 6.7 grammes/tkm. The lubricating oil consumption was 4.16 kg, i.e. 0.37 g/tkm.

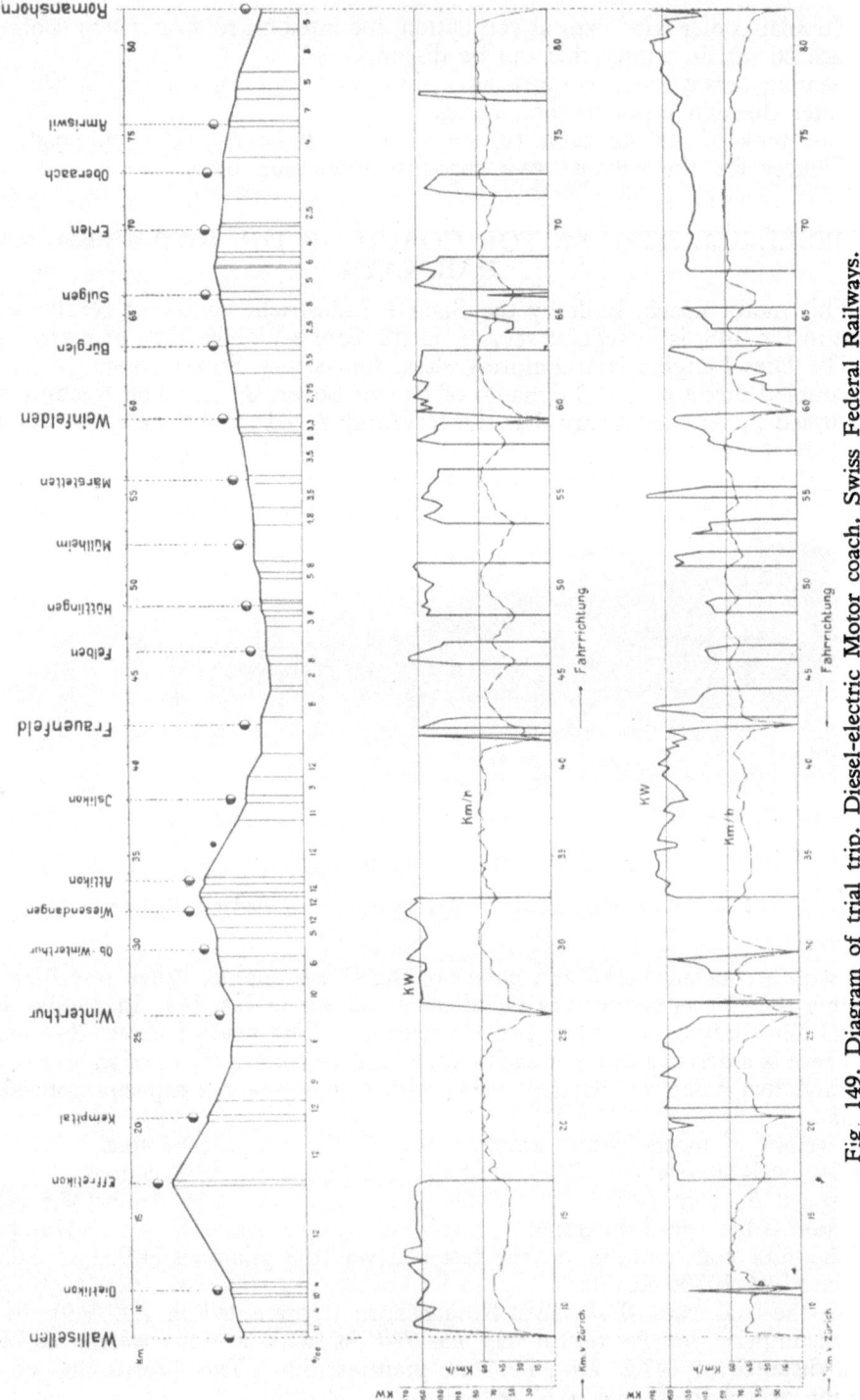

Fig. 149. Diagram of trial trip. Diesel-electric Motor coach, Swiss Federal Railways. Fahrrichtung = Direction of running. — Km v Zürich = km from Zurich.

DIESEL-ELECTRIC MOTOR COACH OF THE APPENZELLERN RLY (SWITZERLAND).

This motor coach, shown in figs. 150 and 151, is a four-axled vehicle for narrow gauge. The Diesel engines were built by Sulzer A.G. of Winterthur, the electric equipment was supplied by the "Maschinenfabrik Oerlikon" and the mechanical part by the "Schweizerische Industriegesellschaft" of Neuhausen.

Fig. 150. Diesel-electric motor coach, Appenzell Railway (Switzerland).

The line on which these motor coaches are in service is 26 km long, with gradients of 37 : 1000 and curves with a radius of 90 metres.

The main dimensions are as follows:

Weight in working order. . . . . 32 tons
Adhesion weight . . . . 16 „
Engine rating at 775 revs./min . . . . 250 E.H.P.
„   „   „ 625   „   200   „
Traction motors. 2
Transmission ratio. 1 : 7.5
Wheel diameter . . . . 722 mm (3' 4$^3/_8$")
Seating capacity 2nd class 6
„   „   3rd   „   24
Luggage space (with 10 emergency seats) . . . . 5.2 m²
Tractive force on driving { max. . . . . 5000 kg (11000 lbs.)
   wheel periphery   { at 25 km/h 2000   „   (4400 „)

Three four-axled trailing coaches with about 150 passengers in all can be hauled uphill, and 8 coaches can be drawn downhill, for which purpose the braking power has been made especially strong.

The Diesel engine is a four-stroke, six-cylinder, Sulzer engine with pre-ignition chamber and is coupled direct to an Oerlikon DC generator. When travelling downhill the Diesel engine can be switched off, as seen in the diagram, fig. 153, of the trial run.

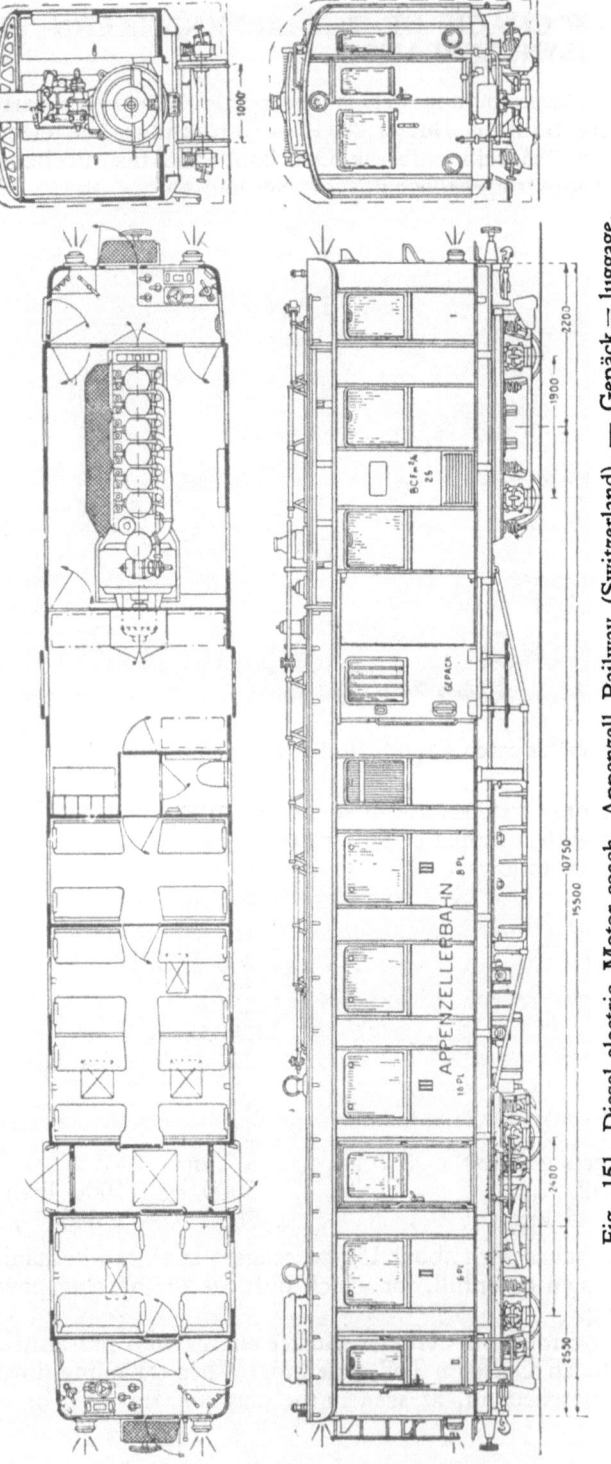

Fig. 151. Diesel-electric Motor coach, Appenzell Railway (Switzerland). — Gepäck = luggage.

The Diesel engine is started by running the generator as motor on current supplied by an accumulator battery. Starting takes only a few seconds and is effected by means of a hand switch in the driver's cabin.

Built on to the main generator driven by the Diesel engine is an auxilliary generator, which supplies current for the field windings of the main generator and for charging the accu. battery.

The wiring arrangement is shown in fig. 152.

The coach comprises two driver's cabins, an engine-room, one 2nd class and two 3rd class compartments, an entrance platform, a lavatory and a luggage compartment. The roof over the engine-room is removable to allow of the engine being lifted out.

In the driver's cabin is a dead-man's control, by means of which the field circuit of the main generator is interrupted and the Westinghouse brake put into action. The braking action does not begin, however, till after a distance of about 50 metres has been covered, so that the driver can leave his post for a moment without the coach being immediately brought to a standstill, which is an advantage when shunting.

In each driver's cabin is a tachometer for the Diesel engine, and an amperemeter and wattmeter for the main generator current. The air compressor for the Westinghouse brake is automatically connected up at a pressure of $5\frac{1}{2}$ atm. and disconnected when the pressure reaches 7.2 atm.

The coach is heated electrically by current from the main generator.

In fig. 153 a diagram is given representing a trial run with one of these motor coaches on the Gossau-Appenzell line. The train consisted of the motor coach with three four-axled trailers, with a total weight of 68 tons. The trial run was made when the motor coach had been in service two months already, and at the time a very strong wind was blowing half-ahead, whilst the rails were covered with about 10 cm snow. The fuel consumption was 13 grammes per tkm, thus rather high, but considering that the trial was on a mountain railway with many curves and long gradients it is not so bad.

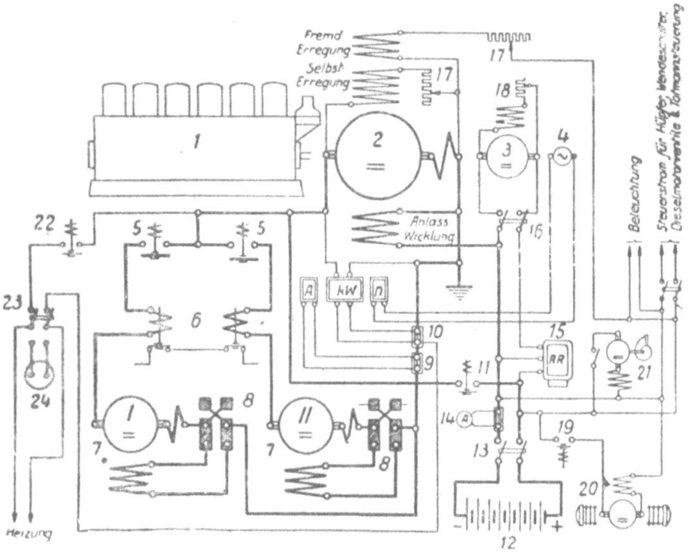

Fig. 152. Wiring diagram.

Anlasswicklung = Coil for starting motor. — Beleuchtung = Lighting. — Fremd Erregung = Exciting. — Heizung = Car heating. — Selbst Erregung = Self-exciting. — Steuerstrom für Hüpfer, Wendeschalter, Dieselmotorventile & Totmannsteuerung = Control current for jump-switch, reverse switch, Diesel engine valves & deadman's control.

1. Diesel engine. 2. Main generator. 3. Auxiliary generator. 4. Tachometer. 5. Main jump-switch. 6. Maximum current relay. 7. Traction motor. 8. Reverse switch. 9. Amperemeter shunt. 10. Wattmeter shunt. 11. Starting jump-switch. 12. Storage battery. 13. Battery switch. 14. Amperemeter for battery. 15. Battery charging relay. 16. Switch for auxiliary generator. 17. Rheostat for regulating field. 18. Rheostat for starting field. 19. Compressor jump-switch. 20. Motor compressor. 21. Cooling water pump. 22. Jump-switch for car heating. 23. Switch for car heating. 24. Plug box for car heating in depots.

Since these Diesel motor coaches have been put into service in the place of steam locomotives it has been found that the fuel expenses have been considerably reduced. Generally speaking the working results on the Appenzellern railway have proved that Diesel electric motor-coaches are highly suitable for railways such as this with steep gradients in mountainous country.

DIESEL-ELECTRIC MOTOR LUGGAGE VAN OF THE SWISS FEDERAL RAILWAYS.

This motor van, shown in fig. 154, was built by the Société Industrielle Suisse of Neuhausen, the Diesel engine being supplied by Sulzer A.G., Winterthur, and the electric equipment by the "Ateliers de Construction Oerlikon".

The Diesel engine has a rating of 420 H.P. This motor luggage van can draw

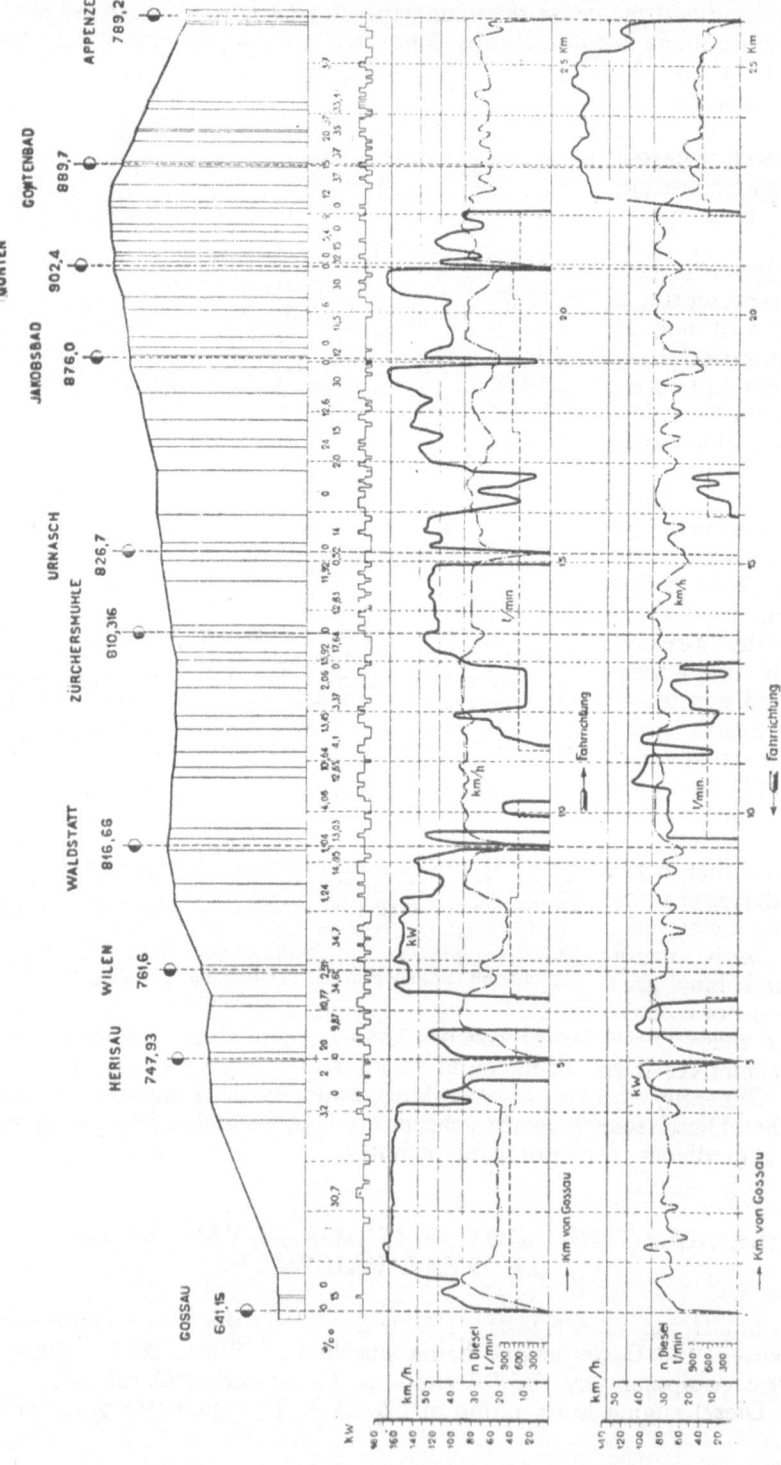

Fig. 153. Diagram of trial trip. Diesel-electric Motor coach, Appenzell Railway (Switzerland). Fahrrichtung = Direction of travel. — km von Gossau = Kilometres from Gossau.

four passenger coaches with a total seating capacity of 200—250. The maximum speed is 75 km/h.

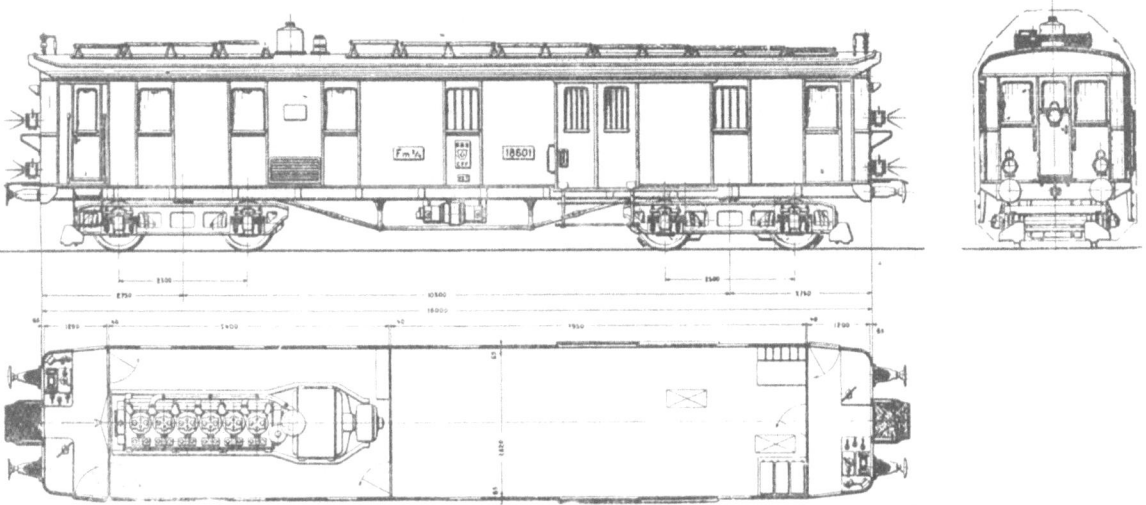

Fig. 154. Diesel-electric Motor luggage van, Swiss Federal Railways.

MOTOR COACHES IN SWEDEN.

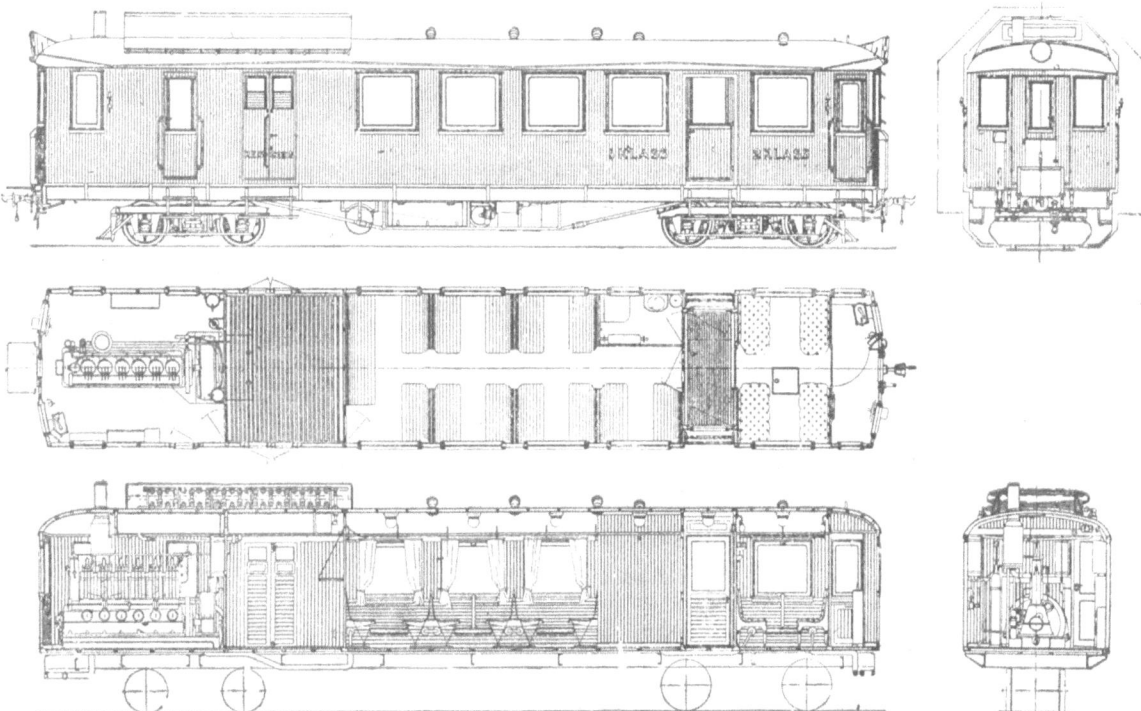

Fig. 155. Diesel-electric Motor coach, Mellersta—Sormlands Railway (Sweden).

Fig. 156. Diesel-electric motor coach, Grangesberg—Oxelosunds Railway (Sweden).

Fig. 157. Diesel-electric motor coach, Boras—Alvesta Railway (Sweden).

The Swedish State Railways have some motor coaches in service fitted with six-cylinder engines of 160—260 H.P., with mechanical transmission. The total weight is 22 tons and seating capacity 60. Without a trailer the motor coach is attended to by one man, but when a trailer is coupled on a second man goes with it.

The fuel consumption is 0.45 litre petrol per km and the oil consumption 15 grammes/km.

In addition to the above, various other railroads in Sweden are using Diesel-electric motor coaches built by "Diesel Elektriska Vagn Aktiebolaget Deva" of Västerås (Sweden), some particulars of which are tabulated below:

Railway	Gauge in metres	Number of motor coaches	Engine rating H.P.	Year taken into service	Kilometres run up to and including 1928	Train service in 1928 train km	Train service in 1928 ton km	Figure
Mellersta—Sormlands	1.435	1	75	1913	833100	88200	2822700	155
Halmstad—Nässjö	"	1	75	1914	905800	56200	1852900	
Skane—Smalands	"	1	75	1915	605200	63900	2025500	
Hälsingborg—Hässleholms	"	1	75	1915	666100	37900	1330600	
Väderstad—Skanninge-Branninge	0.891	1	75	1917	444000	2800		
Norrköping—Soderköping Vikbolandets	"	1	75	1917	350300	49400		
Mellersta—Östergötlands	"	1	75	1923	254400	71500	3030000	
Trafikaktiebolaget Grangesberg—Oxelosunds	1.435	1	90	1925	143500	25200		156
	"	1	90	1926	151700	76400		
	"	2	90	1927	201500	126500		157
Boràs—Alveste	"	1	60	1925	124700	45200		
	"	2	90	1925	260900	86800		
	"	1	90	1927	74300	50000		
Tideholms	"	1	90	1925	181000	55000	1784300	
Norra Södermanlands	"	1	60	1924	217300			
Stockholm—Nynas	"	2	90	1928	83700	83700	3069200	
	"	2	120	1929				

MOTOR COACHES OF THE DANISH STATE RAILWAYS.

In Denmark there are several motor coaches in service, the majority of which are run with petrol engines and some with Diesel engines. The main dimensions are as follows:

	Diesel electric types	Petrol engine types	
Tare	42—44 tons	19.3 tons	11 tons
Axles	4	3	2
Length over buffers........	19.01 m (62′ 4½″)	14.40 m (47′ 3″)	10.72 m (35′ 2″)
Total wheel base..........	15.23 m (50′ 0″)	8.20 m (26′ 11″)	5.90 m (19′ 4¼″)
Seating capacity	59	54	34
Max. speed	70 km/h (44 m.p.h.)	70 km/h (44 m.p.h.)	70 km/h (44 m.p.h.)
Fuel cons. per km.........	600 g	400 g	400 g

As a rule these motor coaches are served by one man, and each coach covers in a year about 270,000 km.

MOTOR COACHES OF THE FINNISH STATE RAILWAYS.

In Finland there are two two-axled petrol motor coaches and one four-axled Diesel motor coach in service, with the following main dimensions:

	Petrol motor coaches	Diesel motor coach
Tare	20.5 tons	33.2 tons
Engine rating	75 H.P.	90 H.P.
Seating capacity		63
Length over buffers		16.35 m (53′ 7″)
Total wheel base......................		10.43 ,, (34′ 2⅝)
Gauge	1.524 m	1.524 ,, (5′ 0″)

The Diesel motor coach was built by the "Diesel Elektriska Vagn Aktiebolaget", Västerås (Sweden).

MOTOR COACHES OF THE CZECHO-SLOVAKIAN STATE RAILWAYS.

In Czecho-Slovakia there are some petrol motor coaches in use, most of them four-axled, and a large number of rail-autobuses on two axles.

A description is given of a couple of types of these vehicles built by the "Tatra-Werke, Automobil und Waggonbau A.G." at Prague-Smichow.

Figs. 158 and 159 show respectively the exterior and the interior of a 4-axled motor coach 3rd class, the main dimensions of which are:

Length over buffers . . . . 21 m (68' 10³/₄")
Length of coach body . . . . 19.80 m (64' 11½")
Distance between bogie centres . . . . 13 m (42' 7³/₄")
Tare . . . . 39 tons
Number of seats 68
Number of folding seats 14
Total number of seats 82
Max. speed . . . . 80 km/h (50 m.p.h.)

Petrol engines
- number 2
- number of cylinders per engine 6
- cylinder bore . . . . 115 mm
- stroke . . . . 180 mm
- revs./min 1200
- rating per engine . . . . 100 H.P.

The vehicle has a driver's cabin at each end, two passenger compartments, an entrance platform and lavatory in the middle and a luggage compartment between one of the driver's cabins and the passenger compartment. Folding seats are fitted on the platform and in the luggage compartment.

In each bogie there is an engine which drives the innermost axles.

The longitudinal and cross girders of the bogie are extended to the end of the car and designed in such a way that with the engine mounted in the bogie the vibrations of the engine are not directly transmitted to the coach body.

The axles rotate is S K F roller bearings.

The motor coaches have Knorr air-brakes, the brake cylinders being placed in each bogie and acting on two brake-drums. The hand screw-brake acts on the wheels of the bogie nearest the driver's cabin.

Sand boxes worked with compressed air are in front of the driving wheels.

The draw-gear of the car is not continuous; drawing forces are transmitted by means of rods and two helical springs.

The car body is entirely of steel, covered on the inside with wood. The floor is of wood, covered with cork sheets over which linoleum is laid.

The cars are heated by the exhaust gas of the engine. Lighting is by electricity, the current being generated by a dynamo connected parallel to an accumulator battery. For giving signals, there is an air whistle and an electric typhon. Rotary pumps are placed at each end of the coach for filling the fuel tanks.

The vehicle is driven by two motors, each mounted on a bogie and connected to a gear box in which the change speed gears, the reversing gears and the axle gears are placed together. This gearbox is mounted on the axle and is suspended resiliently from the bogie. The engaging of the change speed gears and of the reversing gears is done by oil under pressure. The mounting of the motor and gears in the bogie has the advantage of allowing easy replacement in case of a breakdown, for only the bogie has to be replaced to make the coach ready for service again.

The engine is started by electricity, but in case of need it may also be done by hand. Ignition is by means of an electromagnetic high tension igniter.

Four coolers placed on the roof provide for the cooling of the engine; the cooling-water is circulated by circulation pumps driven by the engine.

Fig. 158. Petrol motor coach, Czechoslovakian State Railways.

The four speeds used are engaged by means of oil compressed by air pressure controlled from the control stand.

The change speed gears may be operated from either control stand in such a way that both or either of the driving units may be acted upon. The separate gears of the change-speeds are always in mesh and they are engaged by means of friction clutches, the discs of which are fixed to the gearwheels and to the shafts. By using these discs, the main clutch between engine and driving mechanism is eliminated.

The axle is driven by mechanism mounted in the gear box, together with the change speed gears. The inner wheels of each bogie are driven.

The motor coach has two accumulator batteries of 24 volts and 120 Ah each for lighting and for starting.

Each driver's cabin is equipped with the following apparatus:

a) Master control lever, detachable for operating the change speed gears.
b) Gas control lever.
c) Reversing gear lever.
d) Air brake valve handle.
e) Handwheel for handbrake.
f) Handle for the sanders.
g) Tachometer (A.T. Tachometer and registering tachometer, system Haushälter Řezny).
h) Electric revolution counter; one for each engine in each operating cabin.
i) Double manometer for the air-brake.
k) Lighting switches.

Vehicles as above described, too, exist with 2nd and 3rd class compartments.

One of these motor coaches has electric transmission. Here, too, there are two engines suspended in the car frame; one is mounted in front, the other in the rear, and both are connected rigidly to the dynamos which they drive. The dynamo, too, is mounted on the car frame; the current passes through the change

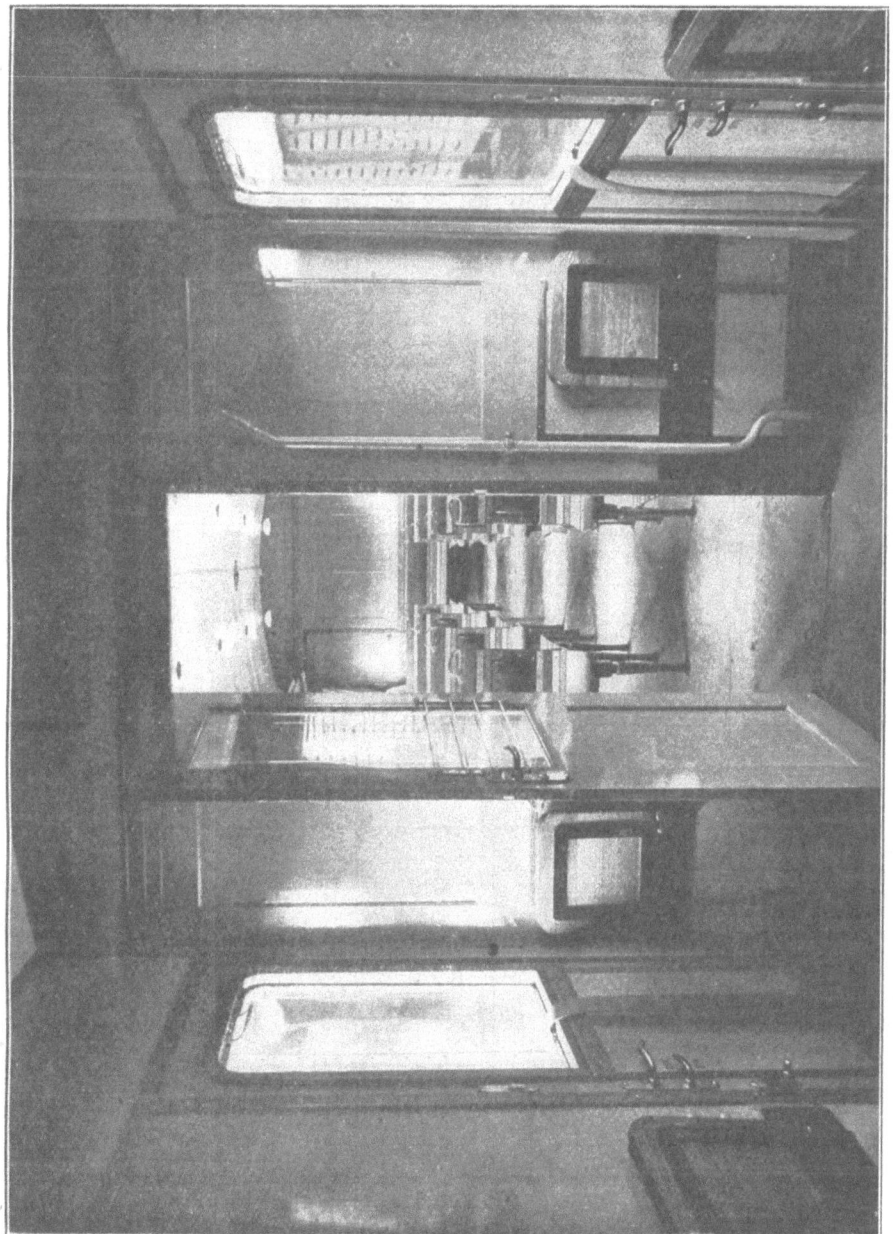

Fig. 159. Petrol motor coach, Czechoslovakian State Railways. Interior.

speed and reversing controllers and so reaches the motors mounted on the driving axles and suspended resiliently in the frame. The inner axles are driven, the outer ones being trailers. The electric arrangement is on the *Gebus* system. Engine and dynamo are mounted on the frame in such a way as to have no contact with metal; rubber plates prevent the transmission of the motor vibrations to the coach body.

Speeds are not changed by gears but by regulating the speed of the engine.

The speed of the car is governed by the generated current and the resistance of the coach. The generated current in turn is governed by the max. rating of the engine or by its number of revolutions (1200), which should not be exceeded; by control of the gas supply this demand is easily satisfied.

Reversing can only be done when the gas handle is in the neutral position.

The car ends are marked P (Přední = front) and Z (Zadni = rear) and the dynamos and electric motors belonging to those parts are marked in the same way. The normal connection is thus: dynamo P with electric motor P and dynamo Z with electric motor Z.

The following combinations may be made by hand:
1. dynamo P with motor P; dynamo Z with motor Z;
2.   „   P „   „   Z; „   Z „   „   P;
3.   „   P „   „   P and Z;
4.   „   Z „   „   P and Z.

From each control stand petrol engine P or Z may be cut out and at the same time the corresponding dynamos and electric motors. This is done in the following cases:
1. when there is an excess of power and
2. in case of breakdowns.

There are two types of two-axled motor coaches, viz. with one and with two driver's cabins.

The main dimensions are given in the table below:

	with one cabin fig. 160	with two cabins fig. 163
Length over buffers....................	9.30 m (30′ 6″)	9.30 m (30′ 6″)
Length of coach body.................	8.50 m (27′ 10³/₄″)	8.50 m (27′ 10³/₄″)
Number of seats......................	36	32
Max. speed	54 km/h (34 m.p.h.)	54 km/h (34 m.p.h.)
Petrol consumption	30 kg/100 km	30 kg/100 km
Weight	8.3 t	
Wheel diameter......................	850 mm (2′ 9½″)	850 mm (2′ 9½″)
Gauge	1435 mm (4′ 8½″)	1435 mm (4′ 8½″)
Petrol-engine { number of cylinders	6	6
bore	90 mm (3⁹/₁₆″)	90 mm (3⁹/₁₆″)
stroke	140 mm (5½″)	140 mm (5½″)
revs./min	1800	1800
rating	65 H.P.	65 H.P.

These motor coaches have central draw and buffer gear.

Figs. 160, 161 and 162 show a motor coach with one driver's cabin; fig. 163 a motor coach with two cabins.

The arrangement of both types is pretty well the same, except for the control stands.

The trailing coaches have a weight of 5.7 tons and a seating capacity for 26 passengers.

The axles are made of chrome nickel steel and rotate in S K F roller bearings.

The Knorr air-brake acts on brake discs on both axles (internal expanding brake). The hand brake acts on the wheels by means of brake shoes. The sanders are worked by compressed air and blow sand in front of the driving axles.

At the ends of the car there are gangways with railings for crossing to the trailer.

The cars are heated by means of the exhaust gas of the engine.

Fig. 160. Petrol motor coach, Czechoslovakian State Railways.

In the cars with one control stand the driver's cabin is in the centre of the car with the driver's seat above the roof.

The cars are lighted by electricity, the current being generated by a dynamo driven by the engine and connected to an accumulator battery.

The car is equipped with an air-whistle.

In the driver's cabin there are the following handles and instruments:

a) Gas handle.
b) Driver's brake valve.
c) Double manometer for the air brake.
d) Speed gauge for indicating the car speed in km/h.
e) Revolution counter.
f) Valve for the sander.
g) Lamp for lighting the control table.
h) Lamp for fuel control.
i) Lamp for control of dynamo for lighting.

Fig. 161. Petrol motor train, Czechoslovakian State Railways.

k) and *l*) Switches for the reflectors and for the signal lamps.

m) Push button for short-circuiting the motor control circuit.

n) Push button for the electric starter.

p) Change-speed gear handle.

r) Clutch pedal.

s) Reversing gear handle.

t) Crank for the hand-brake with pawl.

The coolers for recooling the water from the cylinder jackets are on top of the roof. In the pipe-line between engine and cooler there is a thermostat which functions in such a way that the pipe-line opens when the water has reached a temperature of 80°C. This ensures the temperature of the cooling water being kept constant, at about 80° C.

The coach is driven by an engine placed in the middle of the car underneath the driver's cabin. It acts on a universal joint which transmits its rotation to the axle via bevel gears, which act at the same time as reversing gear.

There are 4 speeds, in the first three of which only two gear wheels are involved.

The compressor for the air-brake is mounted in front of the engine and is driven by gear wheels; it has a max. number of revolutions of 900 and delivers 12.6 m³ per hour at 4.5 atm.

The accumulator has a capacity of 12 volts and 175 Ah.

The motor coaches described here make daily runs of an average of 300 km and about 60% of them are in active service.

The four-axled motor coaches cover yearly about 50,000 km and the two-axled cars (rail buses) about 40,000 km.

Fig. 162. Petrol motor coach, Czechoslovakian State Railways. Driver's cabin.

MOTOR COACHES OF THE ROUMANIAN STATE RAILWAYS.

The Roumanian railways have been using petrol motor coaches with electric transmission for a number of years already.

Fig. 163. Petrol motor coach and trailer, Czechoslovakian State Railways.

The main demensions are:

Axles 2
Engine rating . 60—90 H.P.
Max. speed . . 65 km/h (40 m.p.h.)
Seating capacity. 36
Petrol consump-
tion . . . . . 585—715 g per km
(0.8—1.0 lbs./mile)

The distance covered by each of these motor coaches in a year is about 50,000 km (31,000 miles).

MOTOR COACHES OF THE NORWEGIAN STATE RAILWAYS.

On these railways two-axled and four-axled petrol motor coaches with mechanical transmission are in service, covering 30,000 to 50,000 km in a year. Both types have six cylinder engines and a maximum speed of 55 km/h (35 m.p.h.). The engine ratings of the two-axled vehicles are 75 and 100 H.P., whilst that of the four-axled vehicle is 160 H.P.

MOTOR COACHES OF THE SOUTHERN RAILWAY (ENGLAND).

These coaches are driven by a Drewry combustion engine and are used for branch line service. They have two axles and weigh 8 tons. The seating capacity is 25 and there is also a small luggage compartment.

The petrol engine is of 45—50 H.P. and is mounted at one end of the coach. Transmission is by means of chains and gear wheels. For each direction there are three speeds, viz. 9.6, 19.2 and 40 km/h.

A driver's cabin is placed at each end.

MOTOR COACHES OF THE J. G. BRILL COMPANY OF PHILADELPHIA, U.S.A.

This firm has built a large number of motor coaches for various Ameri-

Fig. 164. Brill driving bogie.

can, Australian and South-African railways, all of which are driven by an explosion engine with electric or mechanical transmission.

The motor coaches with electric transmission have an engine rating varying between 250 and 600 H.P., whilst those with mechanical transmission are the models 75 and 55, the former with a six-cylinder engine of 190 H.P. at 1300 revs./min and the latter with an engine rating up to 175 H.P.

The mechanical transmission is arranged for five speeds with reversing gear, so that all five can be used in either direction; the first two speeds are for starting and the other three for running.

BRILL MODEL 75 PETROL CAR.

This model has a six-cylinder engine of 190 H.P. at 1300 revs./min with mechanical transmission, both axles of the bogie being driven. The engine is mounted at one end of the car, while the transmission gear, with 5 speeds, is placed in the bolster of the engine bogie; all gear wheels are continuously in mesh. Fig. 164 shows the engine bogie and fig. 165 the trailer bogie, while the driving gear is shown in fig. 166. In both bogies bearing springs are provided underneath the journal boxes and the bolster is supported on swing links.

The driving wheels are fixed to the axles by means of tapered hubs in two halves locked by a nut. Thus the wheels can easily be removed when it is necessary to take off the gear wheel box from the axle. The axles rotate in roller-bearings.

The engine is a six-cylin-

Fig. 165. Brill trailing bogie.

Fig. 166 Power assembly of Brill Model 75 Petrol Motor coach.

Fig. 167. Brill model 75 Petrol mechanical drive coach, South Australian Railways.

der, four-stroke explosion engine with 6″ (152 mm) bore and 7″ (178 mm) stroke; its ratings are 75, 150, 190 and 206 H.P. at 500, 1000, 1300 and 1500 revs./min respectively. Carburettors, igniters and vacuum fuel feed are all in duplicate.

The main dimensions of the car are:

Length overall . . . . 55′ (16.764 m)
Length of luggage compartment, containing the engine . . . 16′ 3″ (4.953 m)
Length of passenger compartment . . . . 34′ (10.363 m)
Length of vestibule . . . . 4′ (1.219 m)
Inside width . . . . 9′ 2″ (2.794 m)
Tare . . . . 53000 lbs. (24 tons)

Fig. 168. Brill motor coach, South Australian Railways. Passenger compartment.

The car is lighted by electricity supplied by an accumulator battery.

In trial runs without trailer a speed of 25 m.p.h. (40 km/h) was reached with an average acceleration of 1 mile/hour/sec (0.45 m/sec/sec).

At 1300 engine revolutions per minute the speed of the car is:

in first gear . . . . 59 m.p.h. (95 km/h)
in second gear . . . . 49 ,, (78.5 ,,)
in third gear . . . . 24 ,, (40 ,,)

The motor coach can seat 56 passengers and the trailer 60.

Each cylinder has two inlet and two outlet valves, all in the cylinder head and interchangeable. The advantage of double valves is that they need not be so large, can be better cooled and make less noise. The crankshaft rotates in seven bearings lubricated under high pressure. There are two fuel lines, two fuel tanks, two independent magnetos, two coolers and two cooling water circulating pumps. This duplication of the principle parts enhances the working efficiency. The engine is started by electricity supplied by a generator driven by the engine.

The cooling water is circulated by a double centrifugal pump with a displacement of 50 gallons (225 litres) per minute at 1000 revs./min and 70 gallons (320 litres) at 1300 revs./min. Each half of the pump serves one of the coolers. The total water capacity of the coolers is 30 gallons (136 litres).

Fig. 169. Brill model 75 motor coach, South Australian Railways. Luggage compartment looking towards the passenger compartment.

Figs. 167—170 give an idea of the outer appearance and internal arrangement of such a car supplied to the South Australian Railways.

BRILL MODEL 55 PETROL CAR.

This model is especially designed for branch lines, and such a car (see figs. 171 and 172) is in service on the Gulf, Texas, and Western Railroad.

During a period of 3 months, in which time 19872 miles (32000 km) were covered, the costs were 15.3 dollar cents per train mile and the fuel consumption 0.14 gallons per mile (0.4 litre per km).

Fig. 170. Brill model 75 Motor coach, South Australian Railways. Luggage Compartment looking towards the driver's cabin.

The main dimensions of this car are:

Length . . . . 43′ 5⁵/₁₆″ (13.240 m)
Width . . . . 8′4″ (2.540 m)
Length of luggage compartment. . . . . 11′ 6″ (3.505 m)
Distance between centres of bogies . . . . 22′ 2″ (6.756 m)
Wheel base of bogies . . . . 6′ 8″ (2.032 m)
Wheel diameter . . . . 30″ (762 mm)
Seats in passenger compartment 38
Seats in luggage   „   5
Total weight . . . . 28000 lbs. (12.7 tons)
Four-stroke explosion engine
{ bore . . . . 4³/₄″ (121 mm)
  stroke . . . . 6″ (152 mm)
  normal revs./min 1500
  rating . . . . 68 H.P.
  max. revs./min 1800
Normal speed . . . . 38 m.p.h. (61 km/h)

The transmission from the engine, mounted in a box in the luggage compartment, is via gear wheels, acting on the axles of one of the bogies.

The smallest radius of curve that the car can take is 70 ft. (abt. 21 metres).

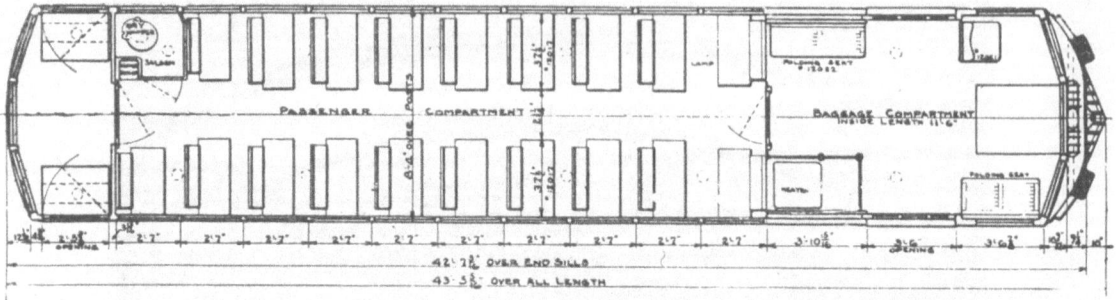

Fig. 171. Brill Model 55 Petrol motor coach, Gulf, Texas and Western Railroad.

Fig. 172. Brill Model 55 Petrol motor coach, Gulf, Texas & Western Railroad.

Fig. 173. Brill Model 30 Petrol motor coach.

BRILL MODEL 30 PASSENGER CAR.

This car is particularly designed for light passenger service. In front the body rests on a bogie and at the rear on a single axle. The main dimensions are:

Length . . . 22′ 8″ (6.909 m)
Width . . . . 7′ 10½″ (2.400 ,,)
Seats 30
Cylinders . . 4
Engine rating 40 H.P.
Normal speed 25 m.p.h. (40 km/h)
Weight in
working order 12500 lbs. (5.6 tons)

Fig. 174. Brill bogie.

Such a car is shown in fig. 173 and the bogie in fig. 174.

BRILL MODEL 43 PETROL CAR.

This car has two compartments, one for passengers and one for luggage. For the rest the design is very similar to that of the Brill Model 30. The main dimensions are:

Length . . . . 32′ 1″ (9.779 m)
Width . . . . 9′ 6″ (2.896 ,,)
Seats in passenger compt. 38
,, ,, luggage ,, ... 5
Cylinders 4
Engine rating . . . . 50 H.P.
Normal speed forwards . . . . 33 m.p.h. (53 km/h)
,, ,, backwards . . . . 9 ,, (14 ,,)
Weight in working order . . . . 22,600 lbs. (10 tons)

This car is shown in fig. 175.

Fig. 175. Brill Model 43 Petrol motor coach.

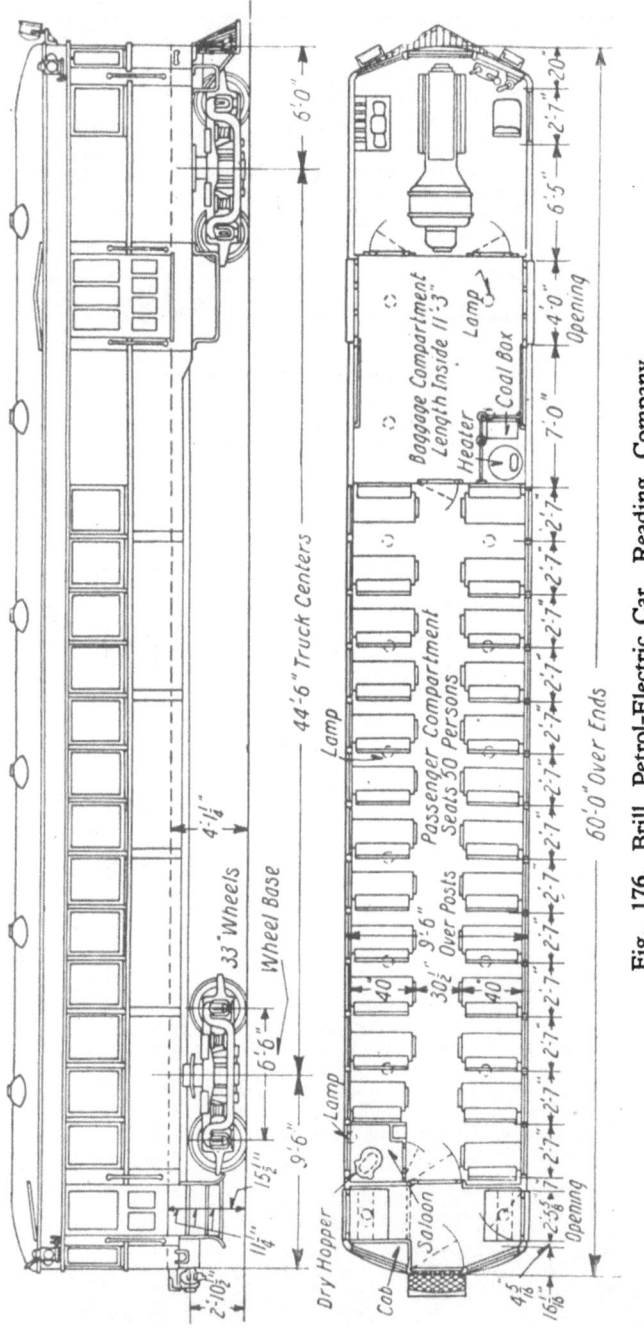

Fig. 176. Brill Petrol-Electric Car, Reading Company.

BRILL's 250 H.P. PETROL-ELECTRIC CAR FOR THE READING Co.

This car (see fig. 176) has a 6-cylinder explosion engine of 250 H.P. with $7\frac{1}{4}''$ (184 mm) bore and 8" (203 mm) stroke, with normal speed of 1100 revs./min. The engine is started electrically. For the drive there are two electric motors of 140 H.P. each mounted on the front bogie.

The driving control is on the Westinghouse electro-pneumatic system, with driver's cabins at both ends.

The main dimensions are:

Length	60' (18.284 m)
Width	9'6" (2.896 „)
Length of luggage compartment	11' 3" (3.429 m)
Wheel base of bogies	6' 6" (1.981 m)
Distance between centres of bogies	44' 6" (13.563 m)
Wheel diameter	33" (838 mm)
Seating capacity	50
Tare 88000 lbs. (40 tons)	

Max. speed without trailer 48 m.p.h. (77 km/h)

Max. speed with trailer load of 150,000 lbs. (67 tons) 36 m.p.h. (58 km/h)

BRILL's PETROL ELECTRIC TRAIN.

The Lehigh Valley Railroad has some motor trains in service as shown in fig. 177. Each train is composed of 4 coaches, one of which is a motor coach with two engines of 250 H.P. each. The motor coach has a luggage compartment in addition to the passenger compartment. The motor coach is 70' 6" (21.488 m) long and can seat 19 passengers, the trailers are 57' 6" (17.526 m) long and can seat 78 passengers.

DIESEL-ELECTRIC MOTOR COACHES OF THE PENNSYLVANIA RAILROAD.

This railroad has two motor coaches in service built by the Pullman Car & Mnfg Corp. of Chicago and fitted with Diesel engines from the South Philadelphia (Pa.) works of the Westinghouse Electric & Mnfg Co., whilst the electric equipment was supplied by the East Pittsburgh Works of the latter company.

The cars are built entirely of steel, are divided into four compartments (nonsmoker, luggage and engine room), and have a closed platform at the rear end, where there is also a driver's stand for running backwards. The control handles etc. are locked up when not in use.

Illustrations of this motor coach are given in figs. 178 and 179.

In the luggage compartment there is a hot water heater for heating all compartments, including also the engine room and the cylinder cooling jackets when the car is not in service. Further the car has electric lighting.

In the front end wall of the coach is a large door through which the engine and all accessories can be passed in and out.

The axles run in Hyatt roller bearings.

The Diesel engine is a four-stroke compressorless six-cylinder engine, coupled direct to the main generator.

The generator is a shunt-wound machine equipped with commutating poles and compound windings for engine-starting and battery-charging at low speed. It is normally operated and separately excited, generator field current being supplied by the auxiliary generator. When delivering power to the traction motors the series winding is cut out of the circuit and external voltage regulation is employed, which acts through the separately excited field. By this method it is possible to make the generator meet the engine characteristics closely throughout the entire operating range.

Ventilation is obtained by means of a fan on the rear of the armature which draws in the air, passing it through the armature and fields. Air inlet and outlet openings in the top half of the frame are protected by expanded metal screens.

The auxiliary generator frame is mounted on the bracket at the commutator end of the main generator. It may readily be removed without disturbing the main generator. The rotor is pressed on to the extension of the main generator shaft, thus eliminating the use of bearings on the auxiliary generator.

The auxiliary generator is a 64 volt, separately excited shunt-wound machine containing a series differential winding which provides the proper characteristic for charging a 64 volt battery.

The traction motors follow the usual standard Westinghouse design for traction service. They are of the series-wound direct current type.

The control of the car has been developed with a view to obtaining simplicity of manipulation by the operator. A master control is used, having two handles, one the main circuit control and the other the reverse lever. These two levers are interlocked mechanically in the usual manner so that the removal of the reverse lever adequately locks the control stand and prevents manipulation by unauthorized persons.

Engine-starting is accomplished by driving the main generator as a series motor

Fig. 177. Brill Petrol-electric train, Lehigh Valley Railroad.

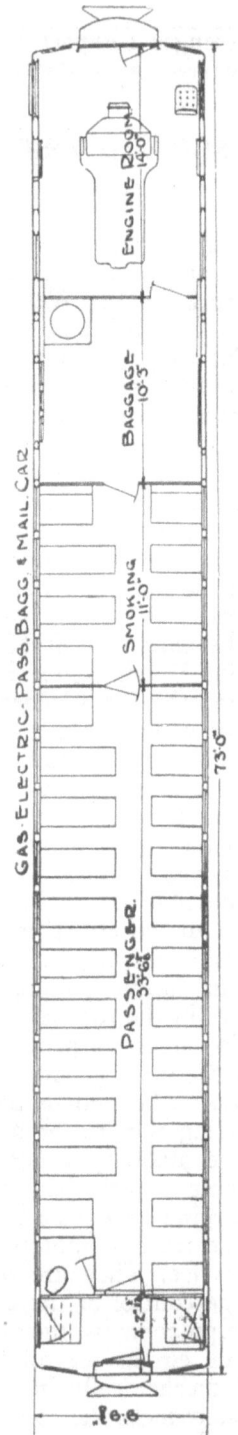

Fig. 178. Westinghouse Petrol-electric railcar, Pennsylvania Railroad.

Fig. 179. Westinghouse Petrol-electric railcar, Pennsylvania Railroad.

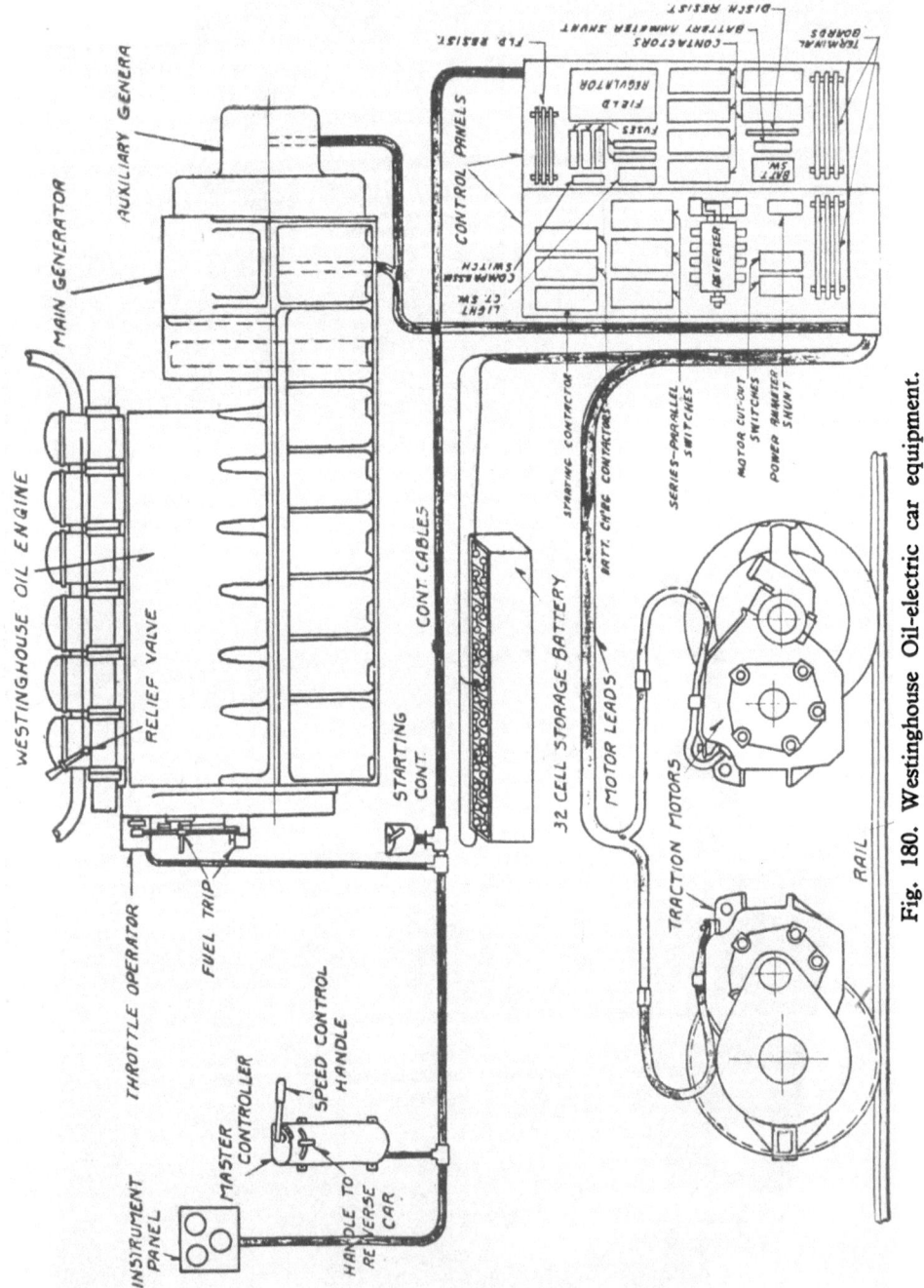

Fig. 180. Westinghouse Oil-electric car equipment.

from the 64 battery. With the engine running the speed is entirely controlled through a system of remote control, making possible the operation in multiple of several engines or railcars.

The power output of the main generator is automatically regulated by the use of a contractor operated by the torque control. This type of control is simple and ensures the delivery to the rail of a great percentage of available engine horse power. At the same time it provides protection against overloading or stalling of the oil engine.

When power is not required for the traction motors the engine may be run at idling speed to reduce fuel consumption. When idling, the main generator furnishes power for air compressor operation and battery charging. With the engine operating at speeds higher than idle, the main generator furnishes power to drive the traction motors while the auxiliary generator supplies power for compressor operation, battery charging and main generator field excitation.

Westinghouse air-brake compressors driven by 64 volt motors are used. The com-

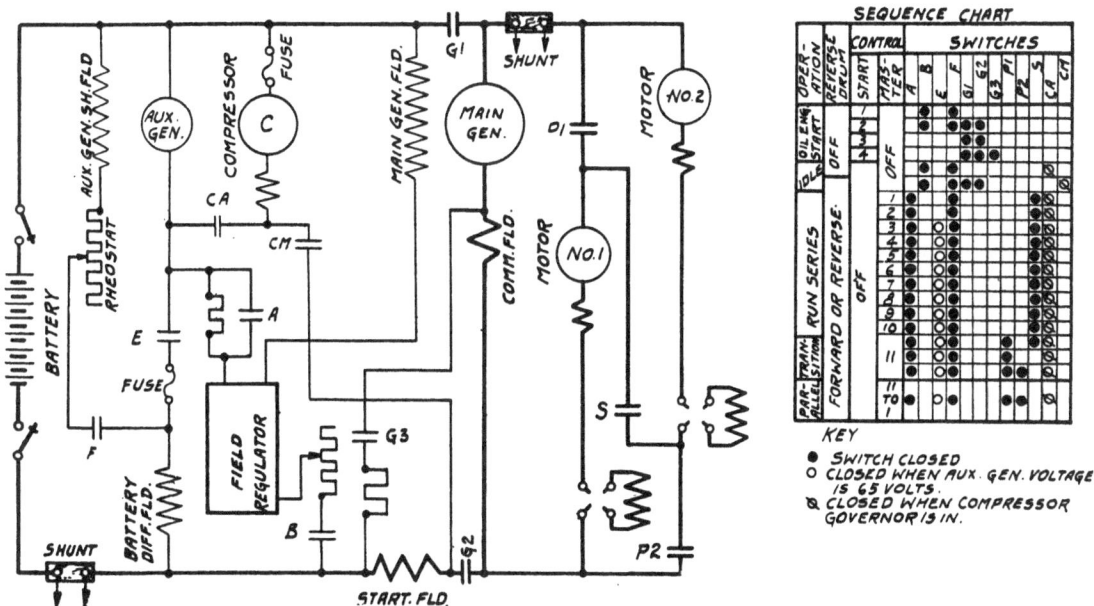

Fig. 181. Westinghouse oil-electric car. Wiring diagram.

pressor control scheme operates automatically to connect them to the main generator circuit during engine idling and transfers them to the auxiliary generator circuit above idle speed. In this way the compressors operate on the normal voltage at all engine speeds.

For starting the railcar the traction motors are connected to the generator in series. The engine speed is gradually raised by notching up on the throttle at the master control. When full engine speed is attained, an additional movement of the control changes the motor connections to parallel by means of closed transition. When field shunting is used, a further movement of the master control handle establishes the traction motor field shunting connections.

Fig. 180 is a diagram of the machinery and fig. 181 the wiring system.

The coach is fitted with a Westinghouse brake with handles in each driver's cabin. For the air-brake there is a dead-man's grip connected up with a foot-pedal in such a way that so long as the driver keeps his foot on the pedal he need not keep hold of the brake handle. The compressor for the air-brake is electrically driven by current from the accumulator battery.

Control from both control stands is on the electro-pneumatic system.

Cooling of the engine cylinders is effected by means of a pump circulating the water through the cooling jackets and return line back to a tank on the roof. The quantity of water in circulation is regulated by a thermostat. From the roof tank the water gravitates through the radiator at the front end of the coach fitted with a screen movable from the driver's cabin, so that it can be let down or raised according to the atmospheric temperature; the position of the screen is indicated in the driver's cabin. The temperature of the cooling water is regulated automatically by the thermostat.

The lubricating oil is cooled by radiators mounted horizontally on the roof. The main dimensions of this motor-coach are:

Max. speed	70 m.p.h.
Max. tractive force	18000 lbs. (8200 kg)
Total length of body	73′ (22.250 m)
,, ,, ,, non-smoking comp.	35′ (10.668 m)
,, ,, ,, smoking compartm.	11′ 10″ (3.607 m)
,, ,, ,, luggage ,,	9′ 6″ (2.896 m)
,, ,, ,, engine room	17′ 6″ (5.334 m)
Width of car body	10′ 4⅝″ (3.166 m)
Wheel diameter	36″ (0.914 m)
Wheelbase motor bogie	7′ 6″ (2.286 m)
,, other ,,	6′ 4″ (1.930 m)

Number of seats	non-smoker	52
	smoker	20
	total	72

Capacity luggage room	5000 lbs. (2250 kg)
Weight on rails of front bogie	85900 lbs. (38.5 t)
,, ,, ,, rear ,,	46900 lbs. (21 t)
Total weight in working order	132800 lbs. (60 t)

Axle journals	front bogie diameter	5½″ (140 mm)
	front bogie length	10″ (254 mm)
	rear bogie diameter	5″ (127 mm)
	rear bogie length	9″ (229 mm)

Diesel engine	rating	300 H.P.
	revolutions p. min	800
	cylinder number	6
	cylinder bore	8¼″ (210 mm)
	cylinder stroke	12″ (305 mm)
Output of generator		210 kW

Traction motors	number	2
	transmission ratio	20 : 59
	hourly rating	350 Amp. at 600 V
	permanent rating	210 ,, ,, 450 ,,

Capacity fuel tank	200 glns. (900 l)

PETROL MOTOR COACH FOR BRAZIL.

This is a two-axled motor saloon coach for 1000 mm gauge and is designed for both adhesion and rack railways. An illustration is given in fig. 182. The coach body was built and fitted out by the "Schweizerische Industriegesellschaft" of Neuhausen. The plan in fig. 183 shows a control stand at each end (g_1 and g_2), with a saloon seating 7 passengers in the middle.

Fig. 182. Mixed Adhesion and Rack Saloon Motor coach Brazilian Railways.

The car is fitted with electric lighting. Special attention has been paid to the heat insulation of the sides and roof in view of the tropical climate in which the car is to be used. On the line in Brazil for which the car is destined some gradients occur

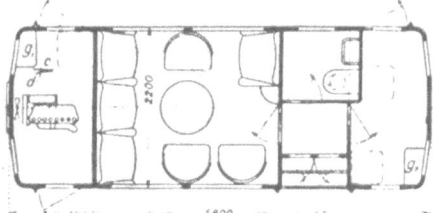

Fig. 183. Mixed Adhesion and Rack Saloon Motor coach, Brazilian Railways.

of 15 : 100, and it is for this reason that it has been designed both for adhesion and for rack tracks.

The engine is a four-cylinder Saurer petrol engine with a rating of 68 H.P. at 1400 revs./min and is mounted in the frame at one end, with the head protruding into the driver's cabin (see fig. 184). In the two end walls there are coolers connected together by pipes. Transmission is on the S.L.M. system, the engine being coupled up by means of a lay shaft and flexible couplings. There are three speeds: 11, 19 and 62 km/h. The reversing gear comprises a dog clutch on the secondary shaft and a lay shaft; the dog clutch is operated by a lever in the driver's cabin. The travelling direction is changed according to

Fig. 184. Saloon Motor coach, Brazilian Railways. Driver's Cabin.

Fig. 185. Saloon Motor coach, Brazilian Railways. Power assembly and transmission.

whether the motor is transmitted via the lay shaft or direct. Mechanical reversing by means of a lever has been chosen because with rail coaches there are no such rapid and frequent reversings as is the case with automobiles. Nevertheless railcars require the same speeds in both directions, and this is easily attainable with mechanical operation. Reversing can only be done while the car is at a standstill.

Light weight being a matter of importance for this coach, the gear box was made of aluminium; it is oil-tight and all rotating parts run in an oil-bath. The axles rotate in ball and roller bearings. The driving gear and engine are rigidly mounted in the frame. Fig. 185 gives a view of the arrangement of the engine, gear box and axle drive; the axle drive consists of a double gear wheel transmission with a number of teeth specially chosen so as to neutralize differences in the diameter of the wheels. The driving gear wheel runs idle on the adhesion sections of the line.

In each control stand is a column with two graduations and two crank handles, the upper one (a) for the engine and engine brake and the lower one (b) for the driving gear. Further there are two brake handles (c and d), one for the brake shoes on the wheels and one for a band brake on the gear wheel brake drums.

The main dimensions of this coach are:

Wheel base . . 2.90 m (9′ 6″)
Length of body 5.78 m (18′ 11½″)
Wheel diameter 724 mm (2′ 4½″)
Tare . . . . 6 tons
Weight in work-
  ing order . 6.8 tons

Trials showed that on gradients not exceeding 37:1000 the car could be run in third gear (60 km/h) and on gradients up to 95 : 1000 in second gear (20 km/h). There was no difficulty in starting in first gear on these gradients, nor on the gradients of 16 : 100 on the Allstätter-Gais Railway. The efficiency of the transmission from the engine to the driving wheel periphery was found to be 90%.

V. DETERMINING THE ENGINE POWER.

The work performed by the engine has to overcome the various forces of resistance occurring during the running of the traction vehicle, with or without trailers. Consequently, in order to decide upon the size of engine required it is necessary to know what these forces of resistance are and how great they are.

In determining the resistance one has to differentiate between that occurring with locomotives, that with motor vehicles and that with trailers, and this resistance is usually expressed as the function of the weight, thus the resistance per ton weight.

The resistances actually occurring are, however, the sum of a number of various kinds of resistance, so that they cannot very well be determined theoretically and can only be fixed experimentally.

Below the results are given of some tests carried out for the determination of these resistances.

According to investigations made by *Lomonossoff* [1]) the resistance of the Russian Diesel electric locomotive unloaded is

$$W = 3 + 0.0015 \ V^2$$

in which W = resistance in kg per ton weight
V = speed in kilometres per hour.

This formula applies only for speeds higher than 8 km/h, the resistance at lower speeds being greater. This resistance is much smaller than that of steam locomotives running with closed throttle, and consequently goods trains drawn by a Diesel electric locomotive can often run down gradients without any work having to be performed, which is not possible with steam locomotives.

The Hungarian State Railways [2]) have determined the resistance of motor vehicles and trailers and thereby arrived at the following formulae, in which the terms W and V have the same meaning as given above.

$$\text{For motor vehicles} \qquad W = 1.77 + \frac{V^2}{550}$$

$$\text{,, trailers} \qquad W = 1.72 + \frac{V^2}{1800}$$

$$\text{,, two-axled goods waggons} \quad W = 2.5 + \frac{V^2}{2000}$$

[1]) "Organ für die Fortschritte des Eisenbahnwesens", No. 7 of 1st April 1928, p. 136.
[2]) ,, ,, ,, ,, ,, ,, Nos. 18/19 of 20th Sept. 1929, p. 332.

From these formulae the following table can be compiled:

Speed in km/h	Resistance in kg per ton weight			
	Diesel-electric locomotive	Motor vehicle	Trailer	Goods waggon
10	3.2	2.1	1.8	2.6
20	3.6	2.5	1.9	2.7
30	4.4	3.4	2.2	3.0
40	5.4	4.7	2.6	3.3
50	6.8	6.3	3.1	3.8
60	8.4	8.3	3.7	4.3
70	10.4	10.7	4.5	5.0

The German State Railways [1]) carried out tests with a four-axled Eva-Maybach Diesel motor vehicle of the following dimensions:

Total length over buffers . . . . 19.3 m (63' 4")
Distance between bogie centres . . . . 11.4 „ (37' 4³/₄")
Length of body . . . . 18.0 „ (59' 0³/₄")

when the resistance on the level was found to be:

speed	10	20	30	40	50	60 km/h
resistance	2.5	2.6	3.2	4.2	5.6	7.1 kg/ton

The Danube Save Adria Railway determined the resistance of their rail-autobuses both loaded and without load [2]) and the results were found to answer the equation

$$W = 1.78 + 0.0029 \, V \left(1 + \frac{F}{G} \, V\right)$$

in which W = resistance in kg per ton weight
V = speed in km/h
F = area of head end of vehicle in sq. metres
G = weight of vehicle in tons.

Taking the average weight of the rail-autobus at 8.2 tons and the area of the head end at 6 sq. metres, the resistance on the level is:

speed	10	20	30	40	50	60 km/h
resistance	1.8	2.7	3.7	5.3	7.2	11.2 kg/ton

For every millimetre size per metre the resistance increases by 1 kg per ton, so that on a gradient of say 1 : 200 the resistance increases by 5 kg per ton weight.

In curves, too, the resistance is increased and according to *Röckl's* formula the increase amounts to:

$$W_b = \frac{650}{R-55} \text{ kg per ton}$$

in which b = curve
R = radius of curve in metres.

---

[1]) "Organ für die Fortschritte des Eisenbahnwesens", No. 2 of 30th Jan. 1926, p. 19.
[2]) „ „ „ „ „ „ No. 2 of 30th Jan. 1926, p. 27.

15

If the engine is rated at N H.P. then at a speed of V km/h it is able to exercise a tractive force of T kg, found from the formula:

$$T = \frac{270\ N}{V}$$

Thus a 75 H.P. engine at a speed of 60 km/h has a tractive force of:

$$\frac{270 \times 75}{60} = 315\ kg,$$

and if this engine is placed in a motor vehicle and a speed of 60 km/h is required on the level the vehicle may not weigh more than:

$$\frac{315}{8.3} = 38\ tons.$$

When this vehicle has to mount a gradient of 1 : 200 an additional resistance of $38 \times 5 = 190$ kg has to be overcome, leaving for propulsion a tractive force of only $315 - 190 = 125$ kg, i.e. per ton weight $\frac{125}{38} = 3.3$ kg, which is not quite sufficient for a speed of 30 km/h. This demonstrates how great an effect a gradient has on the speed.

———

VI. CONCLUSIONS.

As already observed in the foregoing section, three cases have to be distinguished in the application of combustion engines in traction vehicles running on rails, viz.:
a) locomotives of more than 1000 H.P. for main line service;
b) smaller locomotives mainly for shunting work but also for light train service;
c) motor coaches.

Relatively few of the large type of locomotives have as yet been built with combustion engines, and even these few are for the most part experimental, for notwithstanding the energetic efforts undertaken in various countries, the building of Diesel locomotives is still in its infancy. This is most striking in a country like Russia where the thermo-locomotive is certainly in its place in the vast arid regions, the more so considering that the liquid fuel required for the Diesel locomotive is produced in the country itself; it is only in recent years that a few large Diesel locomotives have been put into service, although the subject has been studied for quite a number of years, thanks to the efforts of *Prof. Lomonoss ff*.

Technical development has now reached a stage where highly efficient Diesel engines of 1500 H.P. can be built in a form light enough for use in locomotives. The largest Diesel locomotive as yet put into service (on the Canadian National Railways) has two Diesel engines aggregating 2660 H.P. Steam and electric locomotives, however, are built with a much higher rating and, moreover, these can be temporarily overloaded to a considerable extent, which is not the case with the Diesel locomotive. For the present, therefore, a Diesel locomotive cannot yet be considered for main line express service where such high powers are required.

The following may be mentioned in regard to the efficiency of a Diesel locomotive compared with that of a steam locomotive. With the present modern types of steam locomotives with superheated steam and preheated boiler feed water the consumption of coal is not much less than 1 kg per I.H.P.h. Adding to the thermic losses also the mechanical losses arising in the transmission of the steam power in the cylinder to the periphery of the driving wheels, the total fuel efficiency in the steam locomotive is no more than 8 or 9%. Thus it is not surprising that an improvement was sought in the application of steam condensation to locomotives in combination with the steam turbine, as is applied with success to stationary engines and by means of which steam consumption is reduced to less than 0.5 kg per I.H.P.h. With such a turbo-locomotive an efficiency of $11\frac{1}{2}$—12% can be attained. On the other hand the construction of a turbo-locomotive is expensive and fairly complicated, whilst there are also difficulties connected with the condensation of the steam by air-cooling (the locomotive cannot carry enough water), so that it would seem that for the present the turbine system cannot be applied to any considerable extent to locomotives.

A step is now being taken in another direction by applying steam under very high pressure (60 atm. and more), by means of which the total efficiency can also be increased to 12%. Three types of high pressure locomotive have now been built and are under trial.

Against these figures the total efficiency attainable with a Diesel locomotive with indirect transmission is 25—27%, so that from a point of view of fuel economy the Diesel locomotive is obviously far superior to the steam locomotive either of the present usual type or that with turbo-condensation or high pressure steam. A comparison of the tests carried out at the time with the Russian Diesel-electric locomotive and those on the testbed at Esslingen with a steam locomotive of the same power shows that the average fuel consumption per H.P.h was:

Diesel electric locomotive about 0.24 kg
Steam locomotive   „   0.80 „

the same liquid fuel being used in both cases (calorific value about 10,000 kcal/kg). Thus it is seen that for the same performance the Diesel electric locomotive used only 30% of the fuel required for the steam locomotive.

The saving to be obtained with the Diesel locomotive as compared with a steam locomotive, taking liquid fuel as against coal, can be calculated from the ratio of efficiency and the fuel prices (of oil and coal). These calculations will have to be made for each case separately owing to the fluctuation of prices, both as to time and place, but it must not be forgotten that the fuel prices are to be taken per thermal unit and not by weight. Taking the above-mentioned consumption figures as basis, it appears that when the price of coal per thermal unit is less than one-third of that of Diesel fuel the cost of fuel is less for a steam locomotive than for a Diesel-electric locomotive and conversely when coal costs more than one-third of the price of Diesel fuel the Diesel-locomotive is more economical.

In the case of a railway in a country where oil is produced and coal has to be imported, the conditions may, of course, be more to the advantage of a Diesel locomotive, whilst on the other hand in a country where coal is produced and oil not and freight rates are fairly high, the steam locomotive may be preferable.

A factor in favour of the Diesel locomotive is the absence of the losses inherent to steam locomotives as a result of the boiler having to be fired and steam raised before service is begun, the fire still having to be stoked while the engine is at a standstill, cleaning of the fire after every journey and removing of the fire after the service is ended, a not inconsiderable quantity of useful fuel being thereby lost. A true comparison in this respect can only be made when two locomotives of each kind are run for the same time on exactly the same service, so as to eliminate accidentally favourable or adverse circumstances.

Further, attention is to be drawn to the losses occurring at coal storage depots owing to the breaking up of the coal, weathering and blowing away of the fine dust by the wind. Another point is that stores of liquid fuel can be better guarded than stores of coal in the open.

As a rule the water supply for steam locomotives costs very little as compared with the fuel costs, being generally in the proportion of 1 : 15 or 1 : 20, but in arid regions the water supply may involve considerable expense, and such regions occur in Asia, Australia, Africa, Russia and Central and South America. Again unsatisfactory boiler feed water, causing boiler scale formation, or brackish water may lead to considerable boiler repairs, so that from the point of view of repair costs the Diesel locomotive may have the advantage over the steam locomotive.

As regards the cost of lubricating oil the Diesel locomotive is more expensive than the steam locomotive, but no reliable figures are yet available to allow of a true comparison being made.

The daily upkeep will be less in the case of a Diesel locomotive, for a steam locomotive generally has to have its boiler thoroughly cleaned every 5 or 6 days. For this reason a Diesel locomotive should be able to cover a greater mileage per year than a steam locomotive.

As regards large repairs no definite statement can be made owing to lack of reliable figures in respect of the Diesel locomotive, but considering the periodical boiler and fire-box repairs — after 6 or 7 years all the tubes have to be renewed and after 12—14 years the firebox — a comparison should show to the advantage of the Diesel locomotive.

In respect of weight and initial cost the Diesel compares unfavourably with the steam locomotive. Roughly the weight of the former is $1\frac{1}{2}$ times that of a steam locomotive of the same power, whilst the purchase price is about double.

Turning to the transmission, for the large Diesel locomotives electric transmission is generally the most suitable, as allowing of several axles being driven in a simple manner. On the other hand the heavier and more expensive installation demanded for electric transmission has to be taken into the bargain.

Summarizing, in respect to Diesel locomotives for train service on main lines it may be said that technically they can quite well be adopted, although there is still room for improvement in details. Compared with the steam locomotive a Diesel locomotive has the following advantages:

1) It is always ready for service and involves no losses from the lighting and stopping of fires before and after service. This makes it excellently suitable for use as reserve locomotive on electrified lines, as it can be immediately brought into service if there should be any breakdown of the electric power.
2) There are no fuel losses when the locomotive is standing still, the fuel supply is easily arranged and large supplies can be carried for long journeys.
3) It is eminently suitable for use in arid regions.
4) More mileage can be covered in a year owing to elimination of periodical boiler cleaning and repairs and interruptions for cleaning the fire, bunkering and taking in water.

The disadvantages are the following:
1) The initial cost is about twice that of a steam locomotive.
2) A Diesel locomotive is heavier and more complicated in construction.
3) A special steam boiler may be necessary for heating the train.

Fuel costs may compare in favour of the steam locomotive or in favour of the Diesel locomotive, according to fuel prices.

A reliable prediction cannot easily be made regarding the economic prospects of the Diesel locomotive, there being too many factors, as explained above, which play a decisive part and which can only be judged from experience and application on a larger scale.

After the introduction of the Diesel-electric locomotive on the Russian railways *Prof. Lomonossoff* proceeded to make a calculation of the economic results for a goods-train locomotive [1]). For comparison he took a steam locomotive of the same

---

[1]) „Die Diesel-Electrische Lokomotive" by *Prof. G. Lomonossoff*. Translated from Russian into German by Dr. Ing. Erich Mrongovius, 1924.

power as the Diesel-electric one and determined the cost of transport of one ton useful load per kilometre in so far as such depends on the locomotive. He took into account the following factors:

a) interest and amortization of the locomotive,
b) wages of the crew,
c) maintenance of the locomotive (lighting, lubrication, water supply, etc.)
d) repairs and cost of upkeep in workshops,
e) fuel charges.

Item c he took to be equal in both cases, and item d he considered to be dependent on the quality of the boiler-feed water. Further, he assumed that the purchase price of $ 90,000 for the Diesel-electric is to be compared with $ 50,000 for the steam locomotive, whilst the number of kilometres travelled by the former locomotive in a year will be 50,000 as against 30,000 for the latter.

He thereby arrived at the following conclusions:

1. Profile of the road and nature of the transport have hardly any influence on the saving to be effected with Diesel electric locomotives.
2. If other conditions are equal the amount saved increases with a rise in the price of fuel; with very cheap fuel the oil-fired steam locomotives may even be more advantageous than Diesel-electric locomotives.
3. When fuel is $ 17.— per ton Diesel-electric locomotives are in any case more economical than oil-fired steam locomotives.
4. The advantage of changing over to Diesel-electric traction is all the greater according as the water is less suitable and the number of kilometres to be covered per year is greater.

According to these conditions and at the fuel price quoted, the extent to which costs can be reduced by the use of Diesel-electric locomotives varies from 5 to 40%.

He also calculated that when compared with a steam locomotive fired with Donez coal a Diesel locomotive is the more economical as soon as the price of coal exceeds $ 3.6 per ton.

Prof. Lomonossoff thus concludes that "under these conditions it may be said that at the fuel prices prevailing at the time the change-over to thermo-locomotives on the Russian railways is imperative".

The application of the combustion engine in motor-rail coaches and smaller locomotives and tractors, or locomotors, offers favourable prospects.

In the first place attention may be drawn to the ever-increasing use of small locomotives for shunting work and the so-called locomotors which are so popular for siding and factory work, on account of their easy manipulation and the fact that they are always ready for service. These do not require personnel with several years' training; it takes no longer to learn to drive one of these locomotives than it does to drive a motor car. The perfection of the high-speed Diesel engine so as to enable it to replace the petrol engine will, considering the saving in cost of fuel and elimination of fire risk, do much to promote the adoption of such locomotives. A further consideration is the elimination of boiler upkeep and costly boiler repairs.

For these small locomotives the problem of the transmission from the engine to the driving axles plays no part of importance; as a rule the transmission can be similar to that of an automobile.

These small locomotives are eminently suitable for shunting in small stations where the work to be done is only intermittent, not continuing the whole day long. Usually such a station has not a locomotive depot, so that when shunting has to be done a steam locomotive has to be requisitioned from the nearest depot, and for a large part of the time that it is placed at that station's disposal it is doing nothing, but it still requires firing and watching, so that its crew has to be kept on duty and paid all the time. Where such a station has a locomotor at its disposal it is always ready for service without burning fuel while not in use, whilst the driver can always be given something else to do in between shunting operations. Often it is possible to serve two or three small successive stations with one locomotor, by running it in advance of the goods train and getting the waggons ready for the goods train to pick them up at each station with little delay, thus resulting in quicker transport, quicker return of waggons and locomotives (hence fewer required), and shorter schedule time for a given trip, so that the personnel can be employed for other services. On its return trip, when the goods train has passed the chain of stations, the locomotor can pick up and shunt the waggons left behind by the train into their respective sidings.

Consequently the use of a locomotor for these purposes gives a considerable saving in wages, in fuel and in capital investment; a steam shunting locomotive costs several times more than a locomotor.

As regards the motor-rail coach, this is coming more and more into use everywhere. It is used mainly on lines with little traffic and over not too long distances, serving to connect up several small places with larger main line stations. Motor omnibus competition makes it imperative that better facilities be provided in the place of the notorious local trains (usually composed of rolling stock rejected for main line service but deemed to be good enough for less important branch traffic). Moreover these motor coaches can reach a much higher speed than that attained with the usual local trains; many petrol motor coaches run at 70 km/h, thus much quicker than motor omnibuses. Furthermore they are heated and fitted out much more comfortably and can even be run by one man, as against the three men required for a local steam train.

The dead weight of a local train per seat is much higher than that of a motor coach ($2\frac{1}{2}$—5 times as much), and the hauling to and fro of this dead weight costs power and money, which is one reason why motor coaches can so easily compete with steam trains as regards cost of traction.

As may be seen from various examples given, the use of motor coaches is not restricted to trains comprised of one single motor coach. In a great many cases trailer cars are taken along and often a train is seen with a motor coach at each end and several ordinary coaches in between; with these trains there is the advantage that they can be driven by one man either at the front or at the rear end. Such trains are very similar to electric trains, with the difference that the latter draw their power from overhead cables or a third rail, whereas the Diesel electric trains carry their own power. Although the initial cost of these trains is more than that of an ordinary electric train — in addition to the electric equipment there is also the Diesel machinery — on the other hand there are no electric power stations, sub-stations and overhead cables or third rails to be purchased and kept up, whilst furthermore they can be run on any line and the cost of the current generated is certainly not more — in fact rather less — than that of an electrified line with its inevitable losses — owing to long supply lines and the necessary transformation — and heavy depreciation charges.

This idea of Diesel-electric trains is as yet only in its first stage of development but it already promises much for the future, for the reason that electrification can only be profitable on certain lines with a certain frequency of service and amount of traffic, in view of the high cost of construction.

Thus motor coaches are coming more and more into use, either with simple transmission (gear wheels between engine and driving axles) or some other system such as the hydraulic or pneumatic transmission, or with electric transmission, which latter though expensive is efficient in working and certainly far preferable for short distance traffic.

Dr. Gabriel von Veress, engineer and a managing-director of the Hungarian State Railways, has written an article on the use of motor vehicles in the "Organ f. d. Fortschritte des Eisenbahnwesens", Nos. 18/19 of 20th Sept. 1929, page 371.

Motor vehicles, it is stated, serve only for relatively light traffic, thus low ratings, which cannot be economically served with the rolling stock usually employed on main lines. As such low rating is to be considered that for which the normal steam locomotive cannot be economically used, thus for train loads of about 100 tons and less at average speeds.

On the other hand, in the U.S.A. for instance, motors of 300 to 800 H.P. are used for hauling train loads of much more than 100 tons.

In the "Reichsbahn" of 24th October 1928, on page 922, an article appeared by the State Railway Director *Student*, in which it is stated, in connection with the costs of motor service, that the minimum traffic justifying the use of motor traction is that which the railway company consider to be important enough not to be left to motor-buses. The maximum limit for a "low rating" cannot be fixed, as there are economic conditions playing an important part.

The use of motor traction, however, is not limited by the amount of traffic alone. As is known, the combustion engine is less suitable for overloading than the steam engine. Though not entirely avoidable, traffic fluctuations must be avoided wherever possible on lines with steep gradients. It may be stated as a general rule that the use of motor coaches should be confined to lines with gradients not exceeding 1 : 100, because where steeper gradients occur a certain reserve power is required to overcome them which is not required on the less steep parts of the lines. In this respect high speed Diesel engines have the advantage, as these work with about the same fuel consumption both at half and at full load. On mountain railways, however, the reserve power required would lead to higher capital costs.

If motor traction is used on lines with not too steep gradients there still remain, however, the inevitable regular and irregular traffic fluctuations existing on railways. Temporary fluctuations are due to the following regularly recurring traffic factors:

a) daily fluctuations (workmen's, scholars' and business men's trains to and from the suburbs),

b) weekly traffic (markets and Sunday traffic),

c) annual fluctuations (annual fairs, holidays, sports meetings, etc.).

These fluctuations are known in advance and can easily be coped with.

Irregular fluctuations occur in consequence of sudden changes in the weather (snow, wind and rain) and unexpected crowding, and motor service must be capable of dealing with this heavier traffic as far as possible.

There are three alternative courses to overcome these difficulties:

1) A mixed service can be introduced, heavy trains with steam engines and light trains with motor traction. This, however, involves the maintenance of two kinds of engines, and it is difficult to analyse the running expenses according to the type of traction.

2) The services can be run mainly with motor traction, steam locomotives only being held in reserve for emergencies. The drawback to this system, however, is that the reserve locomotives cannot keep to the schedule time fixed for motor-hauled trains.

3) Motor traction of more than the normally required rating may be used, but in this case the initial cost is higher and running expenses amount to more than is otherwise necessary for normal traffic. If motors of different ratings are in service then the various types again lead to additional cost.

These drawbacks can be reduced somewhat by equipping the smaller and larger motor coaches with engines of the same cylinder units or by coupling two motor coaches together during peak hours.

It is to be noted that against the many disadvantages attaching to motor-coaches with electric transmission there is one great advantage in that they are best suited for load fluctuations, as the combustion engine can always be run with the most favourable number of revolutions.

Compared with steam traction, motor traction has the advantage that with equal track construction greater axle loads and higher speeds are possible. With motor coaches there are no increases or decreases of the rail load, as all moving parts are rotatory; with steam locomotives, on the other hand, such fluctuations do occur, as a result of the slanting position of the driving rod, which gives a vertical component of the driving rod force, and of the balance-weights placed in the driving wheels to partly balance the reciprocating masses.

Further, a motor coach can cover longer distances than a steam locomotive, especially when compared with small local service locomotives, which usually have a relatively small water and fuel carrying capacity.

For economical service it is of great importance that motor traction vehicles should be used judiciously. With proper use the mileage covered per year or between two repair jobs should be as large as possible. When a steam locomotive is under repair the coaches or goods trucks can still be used, but a motor coach is engine and coach in one, so that when this is under repair the income received from its passenger capacity is no longer earned. Consequently motor service is only recommendable where it pays to keep a sufficient stock of engine spares. As far as possible complete engine or machinery units should be kept in reserve for quick replacement, so as to minimize the time during which the motor coach is out of service. If, for instance, a service is run with 6 motor coaches each covering 200 kilometres per day, thus a total of about 430,000 kilometres per year, and one spare engine is held in reserve, then none of the motor coaches need ever be placed out of service for repairs to and overhaul of the machinery, for which about 2 months per year is taken to be required; if there is no reserve engine then there will be one motor coach out of service the whole year round.

An example may be given of a railway district in Hungary served exclusively by motor coaches from the depot at Szentes. Since the introduction of motor traction the number of passengers has increased considerably. At first only mixed trains were run on these lines and the long schedule times gradually led to a busy motor-bus traffic on the roads running parallel to the railways, but in the year 1927/1928 there was an increase in railway passenger traffic of 64.5% as compared with 1925/1926 and 42.6% compared with 1926/27, and some of the bus companies had to suspend their services. The increase in the number of passengers is greater than that of the number of passenger-kilometres, which indicates that the short distance traffic has particularly increased. For these motor services 26 motor coaches are run, covering 4063 km per day, thus per year about 1½ million kilometres. As reserve 4 motor coaches and 2 complete engine sets are kept. With steam traction the same services would require 32 locomotives and 6 in reserve.

Monsieur Paul Beghin, Director of Departmental Railways, Paris, in dealing with the problem of motor coaches [1]) came to the following conclusions:

The petrol motor coach seems to be suitable for light and quick train service on lines with not too steep gradients.

As soon as heavy vehicles have to be used the important advantage of reduction in dead weight per seat is lost and the steam locomotive gains the advantage. Further it is to be noted that the starting of a train of 20 to 25 tons by means of an explosion engine requires special skill on the part of the driver, lack of which is responsible for the damages to the transmission mechanism.

The motor train might be used to replace existing steam trains where there are comparatively few passengers to transport but where there is enough goods traffic to warrant the running of a train every day; if the goods traffic is only sufficient for a train every two or three days, this solution might cause the consignors to get dissatisfied.

The motor coach can also be used as extra train to bring back to rail traffic the passengers lost through the small number of trains placed at their disposal. In order to satisfy the public under such conditions, it is necessary that the route be well chosen so as to pass close to the centres of population, as otherwise the motor bus passing through the centres of the villages will continue to draw the former railway passengers.

The advantages obtained are:

lower costs per kilometre;

higher speed;

no necessity to keep reserve locomotives under steam, with the corresponding saving in personnel; the reserve motor-coach is always ready for immediate service;

fewer out-of-service hours for extensive repairs, provided spare engines are kept and the engines are easily interchangeable;

the cars are cleaner owing to the absence of smoke;

greater comfort for the passengers, particularly if the engine vibrations are absorbed by special arrangements.

The disadvantages are:

high cost of the motor coach, resulting in high charges for amortization, the amount of which cannot easily be fixed on account of the short time that has elapsed since the introduction of motor coaches, it not being possible to say with certainty how long their life will be.

It would be highly desirable if the cost of motor coaches could be reduced by standard manufacture in series, but this does not seem to be likely in the near future, as manufacturers are now applying themselves mostly to the construction of motor-buses, which are very popular just now; this popularity however may disappear when the State and provincial and municipal authorities find themselves burdened with the heavy charges for the upkeep of roads.

Finally, if a fuel more economical than petrol should be found to be of practical utility then the motor coach, even of large capacity, would become of great importance.

---

[1]) "Bulletin de l'Association Internationale du Congrès de Chemins de fer", No. 8, August 1929, p. 1301.

Summarizing, it may be concluded that in every case where traffic has permitted of a steam train being replaced by a motor train, experience has shown that the replacement yields a financial advantage, which will be all the greater when trials with new fuels and even with other sources of energy, such as the electric accumulator, have confirmed the results already obtained.

FINAL CONCLUSIONS.

From the foregoing it appears that for the application of Diesel engines in large locomotives the factors do not apply as exist for automobiles, commercial motor-cars and aeroplanes, for which the steam engine is impracticable, or for ships, where the additional loading capacity offers a considerable advantage.

Saving in fuel is certainly a factor, but in view of the regulation of prices for fuel oil and coal the margin is kept within certain limits.

For motor-trains, small locomotives and tractors there are indeed considerations urging railway companies and factory owners to adopt Diesel traction, and in this direction the saving in fuel costs is naturally a further inducement.

As regards high power locomotives for main railway lines there may be very exceptional cases where Diesel drive is to be recommended, but in most cases the fuel economy attainable will not be sufficient to outweigh the peculiar difficulties accompanying the application of Diesel engines. In the construction of steam locomotives it has always been the aim to have the simplest possible design, and the complication attached to the thermically better systems of steam power — such as the steam turbine with condensation and high boiler pressure — are in this respect certainly more objectionable than those of the Diesel locomotive, so that, considering that its thermic efficiency is twice as great, the Diesel engine promises to take precedence of the improved steam system. It is doubtful, however, whether the saving on the fuel account, which forms but a fraction of the total working expenses of a railway, is to be considered sufficient to set aside the objection of the more complicated construction of Diesel locomotives.

Consequently it would seem advisable to concentrate for the present on the application of the Diesel engine to motor-trains, small locomotives and tractors. Improvements of the high speed Diesel engine will go far to promote its innovation, and in this direction much can still be done.

This would be welcomed by railway and steam-tram companies, who, while realising the importance of Diesel traction, are gradually extending their rolling stock by adding motor vehicles, small locomotives and tractors mostly, for the time being, with petrol engines (which are more expensive in working), because as yet they have not sufficient faith in the high speed Diesel engine.

VII. BIBLIOGRAPHY.

ABBREVIATIONS FOR PERIODICALS.

ENGLISH.

E R — Electric Railway Journal (New York).
Eng — Engineer (London).
Engg — Engineering (London).
L R C — Locomotive Railway Carriage & Wagon Review (London).
M E — Mechanical Engineering (New York).
M T — Modern Transport (London).
P I M E — Proceedings, Institution of Mechanical Engineers (London).
R A — Railway Age (New York).
R E — Railway Engineer (London).
R G — Railway Gazette (London).
R M — Railway Magazine (London).
R M E — Railway Mechanical Engineer (New York).
R R — Railway Review (Chicago).

FRENCH.

B C — Bulletin de l'association internationale du Congrès des Chemins de fer (Brussels).
B S — Bulletin de la Société des ingénieurs civils de France (Paris).
B T — Bulletin technique de la Suisse romande (Vevey).
C T — Les chemins de fer et les tramways (Paris).
G C — Génie civil (Paris).
I V — L'Industrie des voies ferrées et des tramways or L'Industrie des voies ferrées et des transports automobile (Paris).
R G — Revue générale des chemins de fer et des tramways (Paris).
S V — La Science et la Vie (Paris).
R U — Revue universelle des transports et des communications (Paris).

GERMAN.

G A — Glasers Annalen (Berlin).
L — Die Lokomotive (Wien).
Organ — Organ für die Fortschritte des Eisenbahnwesens (München).
S B Z — Schweizerische Bauzeitung (Zürich).
Z V D I — Zeitschrift des Vereins Deutscher Ingenieure (Berlin).
W L B — Der Waggon- und Lokomotiv-Bau (Berlin).
P M K — Der praktische Maschinen-Konstrukteur (Leipzig).

DUTCH.

Ing — De Ingenieur (The Hague).
Loc — De Locomotief (Amsterdam).
S T — Spoor- en Tramwegen (Utrecht).

The articles marked with * are illustrated.

Periodicals				Title	Approx. number of words
Name	No.	Date	Page		
				ENGLISH.	
R M E		24-10-'24	599	New development of the gas-electric car.	1700 *
Eng	3594	14-11-'24	552	A Diesel-electric locomotive.	2800 *
M T	297	22-11-'24	5	A cheap petrol rail motor coach.	900 *
Engg	3076	12-12-'24	817	Petrol-driven rail car with four-wheel drive.	900 *
R A	26	27-12-'24	1155	Brill builds large rail motorcar.	1400 *
R G	3	16- 1-'25	76	400 HP Diesel locomotive.	800 *
R A	5	31- 1-'25	321	STUMPF. Diesel-electric locomotive for the Russian Government.	2200 *
R G	7	13- 2-'25	198	Petrol-driven rail motor vehicles for the Norwegian State Railways.	700 *
R A	7	14- 2-'25	401	Two-car gasoline motortrain for local service on main line.	900 *
R G	8	20- 2-'25	234	The internal-combustion locomotive.	3500
Eng	3613	27-3 -'25	358	An Italian double-ended Diesel-electric locomotive.	1250 *
R R	16	18- 4-'25	741	A new type gas-electric car for the Reading Company.	1200 *
R R	17	25- 4-'25	771	C. & N.W. Ry. converts passenger coach into gas motor car.	2700 *
R A	23	9- 5-'25	1135	Gas-electric car for the New-Haven.	2600 *
Eng	3621	22- 5-'25	578	Diesel locomotive design in Germany.	2000 *
R A	2	11- 7-'25	88	Status of oil-engine locomotives.	3400 *
R M E		July '25	468	Schneider hydraulic transmission for Diesel locomotives.	2800 *
R A	5	1- 8-'25	231	Oil electric locomotive performance.	1500 *
M T	333	1- 8-'25	5	New type petrol-driven locomotive.	800 *
R G	6	7- 8-'25	210	New petrol rail motors, Victorian Government Railways.	300 *
R R	9	29- 8-'25	301	NILSSON. Diesel-electric motor cars are used extensively on Swedish railroads.	4100 *
R A	12	19- 9-'25	507	WANAMAKER. Mc. Keen cars to gas-electric drive.	2700 *
R A	15	10-10-'25	645	Baldwin builds Diesel-electric locomotive.	2000 *
R A	16	17-10-'25	695	Diesel-electric cars for the Canadian National.	3400 *
Eng	3643	23-10-'25	430	Diesel-electric rail cars in Canada.	2200 *
R A	17	24-10-'25	757	Double end controlled Brill gas-electric car.	1000 *
R G	18	30-10-'25	535	Diesel-electric rail cars in Canada.	800 *
R R	19	7-11-'25	689	BAXTER. Canadian National develops oil-electric car.	4400 *
R R	22	28-11-'25	807	BAXTER. New type motor rail car introduced.	2000 *
R M E		November '25	682	Canadian National builds nine motor rail cars.	3800 *
R R	24	12-12-'25	871	Oil-electric versus steam locomotives.	2700 *
R R	26	26-12-'25	939	BOYD. Brill gas-electric car has simple control.	2800 *
R A	26	26-12-'25	1183	New gasoline rail car developed.	1400 *
R A	26	26-12-'25	1190	Oil electric locomotive makes record run.	1900 *
Engg	3131	1- 1-'26	25	Motor rail car developments in the United States.	2500 *

Periodicals				Title	Approx. number of words
Name	No.	Date	Page		
R A	2	9- 1-'26	168	The Reading gas-electric rail car.	2400 *
R A	4	23- 1-'26	273	High capacity gas-electric car for the Seaboard.	1400 *
R G	6	5- 2-'26	189	Internal-combustion locomotive and rail motor-car.	1300 *
R R	6	6- 2-'26	266	Vauclain compares Diesel and steam locomotives.	3800 *
Engg	3137	12- 2-'26	196	250 HP petrol-electric motor rail car.	1200 *
E R	9	27- 2-'26	366	Diesel-electric locomotives discussed in joint New York meeting.	4500 *
R R	9	27- 2-'26	375	Diesel-electric develops great power.	1700 *
R M E		February '26	92	100-ton oil-electric locomotive.	1600 *
R G	10	5- 3-'26	324	Internal combustion locomotives for railway and industrial purposes.	700 *
R A	15	13- 3-'26	809	The Diesel-electric locomotive.	4700 *
R A	17	27- 3-'26	911	NURPHY. Large gas-electric car for the Boston & Maine.	1500 *
R E		March '26	102	Internal-combustion rail car and locomotive.	1400 *
M E		March '26	205	LEMP. Electric transmission for internal combustion engines.	10800 *
R R	18	1- 5-'26	799	Motor cars for branch line economy.	2100 *
R R	19	8- 5-'26	840	GARRISON. Oil electric locomotives show economies.	3500 *
R R	21	22- 5-'26	911	SCARRATT. Automotive train has mechanical drive.	3500 *
R R	22	29- 5-'26	943	PAUL. Gas electric coaches built for C. & A.	2700 *
R M E		May '26	266	VAUCLAIN. Diesel locomotive possibilities.	3000 *
L R C		June '26	178	Petrol auto-cars on the Dutch Railways.	2100 *
M E		June '26	585	The Diesel-electric and its relation to heavy electrification.	6300
Eng	3678	9- 7-'26	34	Geared Diesel locomotive for the Russian State Railways.	2300 *
R A	3	17- 7-'26	103	N.Y., N.H. & H. acquires 73 foot gas-electric car.	500 *
R A	3	17- 7-'26	104	Combination motor car equipped for railway postal service.	800 *
Engg	3159	30- 7-'26	150	Baldwin-Westinghouse 1000 HP Diesel-electric locomotive.	1200 *
R M E		July '26	425	WITTKUHUS. Diesel locomotive with gear transmission.	2900 *
R A	8	21- 8-'26	324	Boston & Maine buys ten more rail motor cars.	1400 *
Eng.	3685	27- 8-'26	230	Petrol rail coaches for Tasmania.	1600 *
M E		August '26	797	LIPETZ. Transmission of power on oil-engine locomotives.	10700 *
M T	391	11- 9-'26	4	New type petrol driven shunting locomotive.	1200 *
L R C		15- 9-'26	281	The Howard petrol locomotive.	800 *
R E		September '26	311	Diesel rail-cars on the Swiss Federal Railways.	2000 *
R G	13	September '26	367	Drewry rail motor-cars for Tasmania.	500 *
R M E		September '26	528	Rail motor cars of all-steel construction.	1600 *

Periodicals				Title	Approx. number of words
Name	No.	Date	Page		
M E		October '26	1043	Transmission of power on oil-engine locomotives.	4500 *
R E		October '26	357	Diesel locomotives for main line traffic.	2700 *
R G	14	October '26	395	Internal-combustion locomotive and passenger car for British Guinea.	500 *
R E		November '26	393	Oil-electric shunting locomotive of 100 tons weight for the Long Island Railroad U.S.A.	1000 *
R G	4	28- 1-'27	112	Large petrol-electric rail coach. Boston- & Maine Railroad.	1000 *
R A	17	26- 3-'27	983	Frisco gets mechanical drive gasoline rail cars.	1500 *
R M E		March '27	146	Gas-electric cars on the D.T. & I.	1400 *
Eng	3718	15- 4-'27	410	KITSON CLARK. Internal combustion locomotives.	3200 *
Engg	3196	15- 4-'27	451	The Kitson-Still internal combustion locomotive.	2500 *
R A	20	16- 4-'27	1195	Rail motor car service on B. & M.	1900 *
R G	18	6- 5-'27	588	Internal-combustion locomotive tractors and inspection cars.	1200 *
Engg	3201	20- 5-'27	608	The development of the Diesel-electric locomotive.	3400 *
R E		May '27	182	Tractive effort of internal-combustion locomotives.	1200 *
M E		Mid-May '27	581	Diesel traction for railroads.	4200 *
R M E		May '27	268	Gasoline rail car for Frisco.	1300 *
R M E		May '27	270	LIPETZ. Diesel locomotive with gear transmission.	4500 *
R M E		May '27	301	Motor rail car service on B. & M.	1800 *
R A	27	4- 6-'27	1739	Diesel cars reduce operating costs on C.N.R.	2500 *
R A	29	18- 6-'27	1939	The Diesel-electric locomotive.	2300 *
R A	3	16- 7-'27	99	ARTHUR. Diesel traction for railroads.	2000 *
R E		June '27	235	New petrol-driven rail motor cars for Spain.	1400 *
R E		August '27	290	The Diesel engine in railway service.	3400 *
P I M E	2	1927	333	CLARK. An internal-combustion locomotive.	23000 *
R A	19	5-11-'27	890	Oil-electric locomotive tested in passenger service.	700 *
R G	20	11-11-'27	585	The first Kitson-Still locomotive completed.	1000
M T	452	12-11-'27	37	Petrol locomotive for shunting operations.	900 *
M T	453	19-11-'27	15	A remarkable internal combustion locomotive.	1500 *
R M E		November '27	719	Diesel-electrics on the Canadian National.	750
Eng	3751	2-12-'27	625	Petrol rail cars for India.	500 *
R A	24	10-12-'27	1165	Rapid fueling facilities for rail motor cars.	400 *
R A	2611	24-12-'27	1305	Rail car or motor coach.	3400 *
R E		December '27	450	Kitson-Still engine completed.	600 *
Engg	3234	6- 1-'28	8	Internal-combustion shunting locomotive.	1100 *
R G	4	27- 1-'28	104	Internal combustion rail motor coaches for India.	1200 *

Periodicals				Title	Approx. number of words
Name	No.	Date	Page		
R M E		February '28	79	HILDEBRAND. Diesel engines for locomotives.	2000 *
R A	9	3- 3-'28	525	Battery-oil-electric locomotive.	2200 *
R A	11	17- 3-'28	635	Diesel engines for railroad traction.	4400
R A	13	31- 3-'28	753	STINEMETZ. What the gas-electric car means to the railroads.	1500 *
R A	15	14- 4-'28	866	Gas-electric unit for railcars.	2600 *
R G	17	27- 4-'28	584	Inspection rail motor for New Zealand Railways.	600 *
Eng	3773	4- 5-'28	484	The Kitson-Still locomotive.	1500 *
Eng	3774	11- 5-'28	514	500 B.H.P. oil electric train.	1700 *
Eng	3776	25- 5-'28	581	Oil burners for Kitson-Stil! locomotive.	1000 *
R M E		May '28	250	Gas-electric unit for rail cars.	2400 *
R A	22	2- 6-'28	1273	GILLILAN. Gas-electric motor cars as applied to steam railroads.	1900 *
M T	482	9- 6-'28	11	FOWLER. Steam and internal combustion locomotives. Recent developments on British and Foreign railways.	1700
R A	23	9- 6-'28	1319	BROOKS. Oil-electric motive power on the Canadian National.	4600 *
R A	24	16- 6-'28	1395	FREEMAN. Reading to use triple-unit gas-electric rail car.	2000 *
R E		June '28	215	Diesel-electric rail motor train, London Midland & Scottish Ry.	2000 *
R M		,, '28	423	Oil engines for rail cars and locomotives.	4600 *
R A	1	7- 7-'28	17	Milwaukee tests Diesel rail car.	1600 *
R G	2	13- 7-'28	53	Internal-combustion locomotive for Australia.	900 *
L R C		15- 8-'28	249	Industrial petrol locomotive: Associated Portland Cement Manufacturers Ltd.	600 *
M T	493	25- 8-'28	3	First oil-electric locomotive built for long haul freight service.	2200 *
M T	493	25- 8-'28	5	Short-distance traffic in Ireland. Petrol rail cars for passengers.	1000 *
Engg	3268	31- 8-'28	265	Oil-electric trains for branch lines.	2000
R E		August '28	303	Latest types of steam and internal-combustion locomotives.	2100
Engg	3275	19-10-'28	496	Diesel-electric rail car for the Pamplona-San Sebastian Railway.	2100 *
R G	16	19-10-'28	490	A new Diesel-electric rail car.	1000 *
R E		October '28	377	A new German Diesel locomotive.	1600 *
M T	504	10-11-'28	13	HARVEY. Oil-electric locomotives in the United States. Long Island Railway development.	2600 *
Eng	3802	23-11-'28	580	A heavy-oil engine locomotive.	1000 *
M T	506	24-11-'28	10	Diesel-engines for light locomotives. The new Kerr Stuart unit.	2700 *
R E		November '28	420	A new Diesel-electric rail car.	2000 *
R A	23	8-12-'28	1125	Canadian National's new 2600 H.P oil-electric locomotive.	2300 *
M T	509	15-12'-28	13	Oil-electric locomotives in Canada. New 2600 HP units for Canadian National Railways.	1000 *
Engg	3287	11- 1-'29	45	Crude-oil engine locomotive.	1600 *
R A	2	12- 1-'29	149	Union Pacific gas-electric cars.	1000 *

Periodicals				Title	Approx. number of words
Name	No.	Date	Page		
R G	4	25- 1-'29	119	An internal-combustion inspection coach.	1100 *
R A	411	26- 1-'29	285	Rock Island finds tractors and trailers valuable.	2000 *
R A	12	23- 3-'29	663	DODD. Diesel-electric passenger locomotive for New York Central.	2700 *
R A	14	6- 4-'29	787	Westinghouse builds 300 hp oil locomotive	2900 *
R A	24	15- 6-'29	1373	HERSHBERGER. Oil-electric rail cars for the Pennsylvania.	3400 *
M T	537	29- 6-'29	6	New Drewry 125 hp petrol locomotive.	1900 *
L R C	443	15- 7-'29	218	Petrol locomotive and transporter wagon, Perak River Hydro-electric Power Company Ltd.	800 *
R A	41	27- 7-'29	251	Rock Island tests powerful rail motor cars.	2500 *
M T	542	3- 8-'29	7	A new Diesel shunting locomotive.	1600 *
R G	7	16- 8-'29	255	Successful Diesel-electric locomotive tests in South-America.	1200 *
M T	547	7- 9-,29	10	Oil-electric locomotives in Canada.	2000 *
Engg	3325	4-10-'29	436	Petrol-driven locomotive for raft transporting wagon.	2200 *
L R C		15-10-'29	334	Economic comparison of the steam locomotive with different types of Diesel locomotives.	1100 *
R A	1711	26-10-'29	1000	Short lines find motor coaches useful.	2300 *
Engg	3331	15-11-'29	627	10-HP petrol-driven locomotive.	3000 *
R G	22	29-11-'29	863	Diesel locomotive development.	
Engg	3334	6-12-'29	748	1200-HP Diesel-compressed-air locomotive for the German State Railways.	600*
R G		7- 2-'30	196	New Diesel compressed-air locomotive.	500*
R E		February '30	81	Three types of locomotives: steam turbine, reciprocating engine and Diesel-electric locomotive efficiences.	3300*
R G		7- 3-'30	343	New 2-6-2 Diesel locomotive for Chile.	1100*
Eng	3869	7- 3-'30	265	A 300 BHP oil engine driven locomotive.	1700*
Engg	3347	7- 3-'30	325	300 HP Diesel locomotive.	1500*
R G		14- 3-'30	394	Diesel-electric traction in suburban work. Buenos Ayres Great Southern Railway.	1400*
Engg		18- 4-'30	504	12-ton petrol locomotive.	1000*
R E	603	April '30	163	New 2-6-2 Diesel locomotive for Chile.	1200*
Eng	3877	2- 5-'30	487	1200 BHP Diesel compressed-air locomotive for the German State Railways.	1800 *
M T		24- 5-'30	13	Oil-electric locomotives for India, to meet road competition.	1500*
Eng	3881	30- 5-'30	600	Oil-electric locomotives for India.	2700*
R E		May '30	200	New Diesel-Electric Locomotive Design.	1000*
Eng	3882	6- 6-'30	361	The Kiston-Still locomotive.	1900*
R A		14- 6-'30		BROWN. An 800- horsepower oil-electric locomotive.	2000*
R E		June '30	233 a. 248	The Diesel-electric system in Rail Service.	1800*
M T		19- 7-'30	5	Steam and Diesel-electric locomotives. Standards of comparison.	1300*
M T		2- 8-'30	3	Steam and Diesel-electric locomotives. Standards of comparison.	3000*

Periodicals				Title	Approx. number of words
Name	No.	Date	Page		
R A		9- 8-'30	280	GEIGER. Diesel Locomotive with compressed air transmission.	1400*
R A		16- 8-'30	326	BREHOB. Oil-electric battery locomotives for the New-York Central.	2300*
M T		23- 8-'30	9	CHORLTON. Diesel-electric traction.	1600
R A		30- 8-'30	451	LIPETZ. Economics of the oil engine locomotive.	3000*
R E		August 1930	317	Diesel compressed air locomotive German Railways.	3000*
R M E		August 1930	454	PALMER LERCH. Gas-electric rail car maintenance on the Reading.	2400*
M T		6- 9-'30	3	Steam and Diesel-electric locomotives.	1100
R E		September '30	349	GEIGER. Diesel locomotive design.	5000*
R E		October '30	381	Diesel rail motors with change speed gearing.	1800*.
Engg	3384	21-11-'30	645	4-6-2 type Ansaldo Diesel locomotive.	1500*
R E		November '30	425	GEIGER. Diesel locomotive design.	3400*
R A	23	6-12-'30	1225	Rail Motor car halves passenger-train costs.	400*

FRENCH.

B S		Mars—Avril 1925	199	BRILLIé. L'Etat actuel de la traction sur voies ferrées par moteurs à combustion interne.	16700 *
G C	2225	4- 4-'25	325	L'Etat actuel de la traction sur voies ferrées par moteurs à combustion interne.	3300 *
I V		Juillet '25	338	Résumé de la note: „Les locomotives pétroléo et Diesel-électrique en construction en Italie.	2300 *
R G		Août '25	174	Transmission hydraulique Schneider pour locomotives à moteur Diesel.	3000 *
G C	2250	26- 9-'25	275	Automotrice pétroléo-électrique Westinghouse.	800 *
C T		Novembre '24	149	Automotrice à boggie pour voie métrique	1500 *
C T		Novembre '25	172	L'évolution des automotrices à essence 1922—1925.	4300 *
G C	2265	9- 1-'26	40	Le train articulé du Canadian National Railway à automotrice pétroléo-électrique.	1400 *
S V		Mars '26	245	DE BRU. Le réseau de l'Etat français vient de mettre à l'essai de nouvelles automotrices à essence.	1700 *
C T		Mars '26	42	VALAT. La transmission mécanique appliquée à la traction sur voie ferrée par moteurs à combustion interne.	5800 *
R G		Mars '26	197	POLART. Note sur les loco-tracteurs à essence de Paris—Saint Lazare.	2000 *
R G		Mars '26	233	Résultats en service de la locomotive Diesel des Chemins de fer russes.	2700 *
G C	2279	17- 4-'26	349	Nouvelle locomotive pétroléo-électrique Baldwin-Westinghouse.	1500 *

Periodicals				Title	Approx. number of words
Name	No.	Date	Page		
B T	11	22- 5-'26	125	Muller. La nouvelle automotrice à un seul agent du chemin de fer Berne-Worb.	1400 *
C T		Mai '26	75	Spiess. Locomotives pétroléo-électrique.	1000 *
C T		Mai '26	77	Roger. Service d'automotrice de la ligne de Pallet à Vallet.	800 *
G C	2288	19- 6-'26	537	Les autobus pétroléo-électriques à huit roues du Chicago and Alton Railroad.	2500 *
C T		Juin '26	89	Spiess. Automotrices et locomotives Diesel-électriques.	2600 *
B T	16	31- 7-'26	198	Course d'essai d'une voiture automotrice à benzine sur la ligne de la Thurgovie centrale les 7 et 8 avril 1926.	400 *
I V		Juillet '26	311	Poullain. Note sur l'utilisation en Tunisie d'un locotracteur Diesel électrique.	2700
G C	2298	28- 8-'26	169	Nouvelle locomotive à moteur Diesel des chemins de fer russes.	1700 *
C T		Octobre '26	166	Maincent. Nouveau type d'automotrice légère à essence de la Compagnie des tramways départementaux des Deux-Sèvres.	1200 *
C T		Novembre '26	187	Maincent. Les résultats économiques que l'on peut attendre de l'emploi des automotrices sur les grands réseaux.	4700
I V		Avril '27	207	Mellini. Automotrices actionnées par moteurs à huiles lourdes Diesel et analogues.	6000 *
R G		Juillet '27	36	Poullain. Description et essais d'une nouvelle locomotive Diesel électrique en service en Tunisie.	4700 *
C T		Juillet '27	126	Locomotives et automotrices Diesel.	3800 *
C T		Août '27	154	Automotrices à combustion interne.	3300 *
C T		Novembre '27	217	Automotrices Diesel-électriques des Canadian National Railways.	1900
I V		Décembre '27	546	Vergniole. Les automotrices à essence sur les réseaux français d'intérêt local. Etude technique et économique.	10200 *
G C	2369	7- 1-'28	17	Nouvelle locomotive Diesel-électrique en service en Tunisie.	900 *
R G		Janvier '28	59	Essais de la locomotive Diesel à transmission mécanique, du Professeur Lomonossoff.	1900 *
C T		Août '28	185	Locomotive Diesel-électrique de la Compagnie Fermière des Chemins de fer Tunisiens.	2000 *
R U		Août '28	77	Wagner. L'utilisation des locomotives Diesel sur les voies ferrés secondaires.	3300
C T		Septembre '28	197	Spiess. Automotrices Diesel-electriques.	6000 *
I V		Octobre '28	302	Mellini et la Valle. Automotrices sur rails à moteurs à combustion interne. Rapport présenté au XXIe Congrès international (Rome 6—12 mai 1928).	20000 *
G C	2417	8-12-'28	560	Nouvelles locomotives Diesel-électrique pour les chemins de fer russes.	1200

Periodicals				Title	Approx. number of words
Name	No.	Date	Page		
C T		Décembre '28	266	Locomotive à essence des Chemins de fer fédéraux.	900 *
G C	2423	19- 1-'29	59	Les automotrices à moteurs Diesel et à transmission électrique du chemin de fer de Pampelune à Saint-Sébastien.	1500 *
C T		Mars '29	48	SPIESS. Locomotive Diesel-électrique du Canadian National Railway.	3100 *
C T		Mai '29	94	SPIESS. Locomotive à voyageurs Diesel électrique du New York Central Railroad.	3700 *
C T		Juillet '29	146	Considérations générales sur les automotrices à combustion interne et transmission électrique.	3900 *
C T		„ '29	151	Utilisation des locomotives Diesel électriques comme locomotives de manoeuvre.	1300
R G		„ '29	64	Locotracteurs Diesel à transmission électrique.	3000
R G		Août '29	204	Etude comparative de la locomotive à vapeur et des différents types de locomotives Diesel au point de vue économique.	1400
G C	19	9-11-'29	460	BARD. Les voitures automobiles Diesel-électriques du chemin de fer d'Appenzell (Suisse).	3600 *
C T		Novembre '29	220	Résultats obtenus au Canada et en République Argentine avec une locomotive Diesel-électrique.	900 *
C T		Avril '30	64	Locomotive Diesel de 300 HP à transmission hydraulique.	2600*
T E		Avril-Mai '30	86	RUEGGER. Les nouvelles automotrices Diesel-électriques employées sur les chemins de fer suisses.	2200*
G C	2493	24- 5-'30	497	Locomotive Diesel pneumatique de 1200 chevaux des chemins de fer allemands.	2700*
C T		Juillet '30	124	SPIESS. Locomotive Diesel-électriques.	3200*
G C	2499	5- 7-'30	9	Les locomotives Diesel-électriques du „Great Southern Railway" de Buenos-Aires.	1500*
C T		Novembre '30	192	Une nouvelle locomotive „Diesel-air comprimé" pour les chemins de fer allemands.	3100*

GERMAN.

Organ	3	15- 2-'25	39	DRAEGER. Die Triebwagen auf der Seddiner Ausstellung.	8000 *
Z V D I	10	7- 3-'25	321	MEINEKE. Vergleichsversuche zwischen Diesel- und Dampflokomotive.	1000 *
Organ	5	15- 3-'25	77	LOTTER. Die erste Diesel-elektrische Vollbahn Güterzuglokomotive.	3500 *
Organ	5	15- 3-'25	82	UEBELACKER. Vergleichsversuche zwischen der russischen Diesel-elektrischen Lokomotive und der russischen E-Heissdampflokomotive auf dem Prüfstand in Esslingen.	1500 *

Periodicals				Title	Approx. number of words
Name	No.	Date	Page		
Organ	7	15- 4-'25	168	Die erste Diesel-elektrische Vollbahn-güterzuglokomotive.	600 *
Z V D I	19	9- 5-'25	635	MAYER. Die Diesellokomotive vom Standpunkt des Lokomotivbaues.	3500 *
Z V D I	19	9- 5-'25	642	GEIGER. Dieselmotor und Kraftübertragung für Grossöllokomotiven.	4500 *
Z V D I	19	9- 5-'25	647	SCHUMACHER. Rohöllokomotiven mit Kompressorlosem Dieselmotor und Flüssigkeitsgetriebe.	2000 *
Organ	12	30- 6-'25	247	ACHILLES. Ueber die Ausführung von Diesellokomotiven.	6000 *
Organ	12	30- 6-'25	255	DANNECKER. Die Diesellokomotive auf der Hauptversammlung des Vereins Deutscher Ingenieure in Augsburg.	4000 *
G A	5	1- 9-'25	93	Benzin-elektrische Lokomotiven für Südwest-Afrika.	300 *
Z V D I	42	17-10-'25	1321	MEINEKE. Betriebs- und Versuchsergebnisse der russischen Diesel-elektrischen Lokomotive.	1500
Z V D I	44	31-10-'25	1387	LOMONOSSOFF. Fahrtergebnisse der Diesel-elektrischen Lokomotive in Russland.	3500 *
Organ	2 / 4	30- 1-'26 / 28- 2-'26	19 / 55	EBEL. Die neuen Verbrennungstriebwagen der Deutschen Reichsbahn-Gesellschaft und ihre Versuchsergebnisse.	6000 *
Organ	2	30- 1-'26	23	POGANY. Schienenautobus.	3700 *
Organ	5	15- 3-'26		Die Eisenbahnfahrzeuge auf der Deutschen Verkehrsausstellung, München 1925.	
			89	Oeltriebwagen.	2000 *
			106	Oellokomotiven.	500 *
G A	7	1- 4-'26	113	Probefahrt mit dem Wumag-Oeltriebwagen auf der Strecke Görlitz—Oberschreiberhau.	800 *
Z V D I	14	3- 4-'26	476	LJUBIMOFF. Dieselgetriebelokomotive mit Magnetkupplungen.	500 *
Z V D I	16	17- 4-'26	527	DOBROWOLSKI. Verbrennung von Masut im Dieselmotor der russischen Diesel-elektrischen Lokomotive.	1200
G A	3	1- 8-'26	39	Probefahrt der Diesel-getriebelokomotiven der Hohenzollern A.G. für Lokomotivbau.	1500 *
Ing	47	20-11-'26	Bijvoegsel.	BROWN. Der heutige Stand des Diesellokomotivbaues.	11000*
G A	11	1-12-'26	164	SCHULZ. Ueber Motorlokomotiven.	3700 *
Z V D I	13	19- 3-'27	389	SCHMINKE. Schwedische Lokomotive mit Flüssigkeitskupplung.	2000 *
Z V D I	25 / 27	18- 6-'27 / 2- 7-'27	873 / 959	DOBROWOLSKI. Die Diesel-Getriebelokomotive und ihre Erprobung.	4200 * / 4200 *
Z V D I	30	23- 7-'27	1046	LOMONOSSOFF. Der gegenwärtige Stand des Diesellokomotivbaues.	3000 *
G A	4	15- 8-'27	62	FEIST. Das Oelschaltgetriebe System S. W. und seine Verwendung im Triebwagenbau.	1300 *

Periodicals				Title	Approx. number of words
Name	No.	Date	Page		
Organ	3	15- 2-'27	39	GÜNTHER. Die mechanisch angetriebene Diesellokomotive mit fester Ueberset-zung und mehreren, einzeln kuppel-baren Motoren.	
	15	15- 8-'27	283		5800 *
Organ	12	30- 6-'27	213	NOLDE. Die neuen Verbrennungstrieb-wagen der Deutschen Reichsbahn-Ge-sellschaft und ihre Versuchsergeb-nisse.	2500 *
Organ	22/23	30-11-'27	442	ENGELS. Neuerungen im Waggon- und Triebfahrzeugbau bei den Oesterrei-chischen Bundesbahnen.	700 *
Z V D I	38	17- 9-'27	1329	LOMONOSSOFF. Zur Theorie der Gas-übertragung bei Diesel-lokomotiven.	4000
Z V D I	49	3-12-'27	1710	Die unmittelbar angetriebene Dieselloko-motive.	4500 *
Z V D I	3	21- 1-'28	90	DOBROWOLSKI. Vergleichsversuche mit russischen Diesellokomotiven.	600 *
Organ	6	15- 3-'28	109	NOLDE. Eine neue Triebwagenbauart mit kompressorlosem Dieselmotor und ihre Versuchsergebnisse.	2400 *
Organ	7	1- 4-'28	133	LOMONOSSOFF. Widerstand und Trägheit der Diesel-elektrischen Lokomotive.	2500 *
Z V D I	17	28- 4-'28	557	SÜBERKRÜB. Die Steuerung Dieselelek-trischer Lokomotiven.	5300 *
Organ	9	1- 5-'28	164	JUDTMANN. Die neuzeitlichen Triebwa-gen Nordamerikas.	7500 *
Z V D I	18	5- 5-'28	603	VETTER. Diesellokomotive mit Flüssig-keitsgetriebe, Bauart Schwartzkopff-Huwiler.	1500 *
Z V D I	21	26- 5-'28	714	MEINEKE. Still-Diesellokomotive von Kit-son & Co., Leeds.	500 *
G A	2	15- 7-'28	20	ACHILLES. Lokomotiven mit Antrieb durch Oelmotor und Dampfmaschine.	2600 *
Organ	19	1-10-'28	416	LOMONOSSOFF. Das Lokomotivstufenge-triebe.	4800 *
Z V D I	44	8-11-'28	1605	NASKE. Neuere Oeltriebwagen.	4500 *
G A	11	1-12-'28	146	GERSTMEYER. Die Diesel-Lokomotive und die moderne Zugförderung.	5800 *
Organ	8	15- 4-'29	124	STRASSER. Die Wirtschaftlichkeit der Die-sel-Lokomotive im Vollbahnbetrieb.	7500 *
	9	1- 5-'29	143		5000 *
W L B	8	18- 4-'29	113	WITTE. Der Entwicklungsgang der Die-sellokomotive, insbesondere derjeni-gen mit Druckluftübertragung bei der Deutschen Reichsbahn.	5000 *
	9	2- 5-'29	129		2500 *
	10	16- 5-'29	145		12000 *
	11	30- 5-'29	163		2500 *
W L B	8	18- 4-'29	123	Die Unterbringung der Maschinenanlage bei Oeltriebwagen.	1300 *
G A	1245	1- 5-'29	135	MÜLLER. Die Diesellokomotive der Tune-sischen Eisenbahngesellschaft.	2600 *
G A	4	15- 8-'29	57	MANGOLD. Personenzug-Diesel-lokomo-tive 2-4-2 für das russische Profil.	4300 *
W L B	16	8- 8-'29		SPIESS. Verbrennungs-Triebwagen und Triebwagenzüge mit elektrischer Kraft-übertragung.	
			241		3500 *

Periodicals				Title	Approx. number of words
Name	No.	Date	Page		
G A	1	1- 7-'29	1	MANGOLD. Personenzug-Diesel-lokomotive 2-4-2 für das russische Profil.	3000 *
Organ	18/19	20- 9-'29	330	HIRSCHMANN. Förderung des Personenverkehrs auf den Lokalbahnen durch Triebwagen und Kleinlokomotiven.	2800 *
Organ	18/19	20- 9-'29	371	VON VERESS. Grundsätzliches über die Verwendung von Oeltriebwagen.	2200 *
W L B	23	14-11-'29	353	WILLIGENS. Disellokomotiven.	2000 *
	24	28-11-'29	369		2500 *
Organ	23	1-12-'29	487	GRÜNING. Diesel-elektrische Lokomotiven.	5500 *
G A	3	1- 2-'30	35	WITTE. Die Fertigstellung der ersten 1200 PS- Diesellokomotive für die Reichsbahn.	3500*
L	2	Februar '30	24	Motortriebwagen der Oesterreichischen Bundesbahnen.	2000*
Z V D I	10	8- 3-'30	289	WITTE & WAGNER. Die 1200-PS-Diesel-Druckluftlokomotive der Deutschen Reichsbahn.	5000*
Z V D I	12	22- 3-'30	366	GEIGER. Diesellokomotieve mit Drucklieftübertragung.	1500 *
W L B	11	5- 6-'30	164	SCHWETER. Diesel-Lokomotiven mit unmittelbarem Antrieb.	1500*
Organ	17	1- 9-'30	379	GEIGER. Ueber Diesellokomotiven mit besonderer Berücksichtigung der Dieseldruckluftlokomotive.	5000 *
Z V D I	50	12-12-'30	1697	NIEDERSTRASSER. Leichte Motor-Verschiebelokomotiven der Deutschen Reichsbahn.	3000*

DUTCH.

Name	No.	Date	Page	Title	Approx. number of words
Ing	2	10- 1-'25	25	HUPKES. Benzine motorrijtuigen der Nederlandsche Spoorwegen.	8000 *
Loc	47	25-11-'25	369	Diesel-electrische motorwagens voor spoorwegverkeer.	2200 *
Loc	15	14- 4-'26	116	De verbrandingsmotor in zijne toepassingen bij de tractie der spoorlijnen.	1700 *
Loc	22	2- 6-'26	172	Een jaar motorwagendienst op de Südstor-marnsche Kreisbahn.	2600
Ing	47	20-11-'26	Bijvoegsel.	BROWN. Der heutige Stand des Diesellokomotivbaues.	11000 *
Loc	19	11- 5-'27	145	Een nieuw onderstel voor motorwagens.	2900 *
Loc	21	25- 5-'27	161	De motorwagens met verbrandingsmotor.	2200 *
Loc	38	19- 9-'28	299	De concurrentie tusschen motorwagens met verbrandingsmotoren en stoomlocomotieven.	1700
Loc	31	31- 7-'29	241	Automobielen tot spoorwegrijtuigen omgebouwd.	400
S T	8	15-10-'29	188	DE GELDER. Olie-electrische locomotieven.	1600 *
Ing	48	30-11-'29	V93	VAN DER ZEE. De exploitatie van den Woldjerspoorweg met benzinemotorrijtuigen.	1500 *

Periodicals				Title	Approx. number of words
Name	No.	Date	Page		
S T	1 2	7- 1-'30 21- 1-'30	4 40	BOLLEMAN KYLSTRA. De nieuwe benzine-motorrijtuigen der Nederlandsche Spoorwegen.	3000 *
S T	2	21- 1-'30	44	LABRIJN. Nieuwe locomotoren voor de Nederlandsche Spoorwegen.	1700 *
S T	7	29- 9-'31	173	LABRIJN. Ombouw der B_o Accumulatoren Locomotieven Nos. 83 en 84 tot B_o Diesel-electrische locomotieven.	1500*

BOOKS.

ENGLISH.

HOBSON. The internal-combustion locomotive. Paper read before the North-East Coast Institution of Engineers and Shipbuilders, Newcastle-upon-Tyne. February, 13, 1925.

LIPETZ. Transmission of power on oil engine locomotives. 1926.

LIPETZ. The present status of the oil engine locomotive. Chicago 1927.

GERMAN.

LOMONOSSOFF. Diesel-Lokomotiven. 1929. V.D.I.-Verlag G. m. b. H. Berlin NW 7.

LOMONOSSOFF. Die Diesel-Elektrische Lokomotive. V.D.I.-Verlag G. m. b. H. Berlin NW 7.

JANKE. Der Eisenbahn-Oeltriebwagen. Leipzig, Oskar Leiner.

BROWN. Ueber Diesel-elektrische Lokomotiven im Vollbahnbetrieb. 1924. Ernst Waldmann Verlag, Zürich.

BAUER. Diesellokomotiven und ihr Antrieb. 1925. Verlag von Julius Springer, Berlin und Wien.

SÜDERKRÜB. Fahrzeug-Getriebe. 1929. Verlag von Julius Springer, Berlin und Wien.

INDEX.